Oh Omnivore!

The Path of an Eclectic Theoretical Biologist

Oh Omnivore!

The Path of an Eclectic Theoretical Biologist

RICHARD (DICK) GORDON

Published by

World Scientific Publishing Co. Pte. Ltd.

5 Toh Tuck Link, Singapore 596224

USA office: 27 Warren Street, Suite 401-402, Hackensack, NJ 07601

UK office: 57 Shelton Street, Covent Garden, London WC2H 9HE

British Library Cataloguing-in-Publication Data
A catalogue record for this book is available from the British Library.

OH OMNIVORE!
The Path of an Eclectic Theoretical Biologist

ISBN 978-981-98-1144-1 (hardcover)
ISBN 978-981-98-1238-7 (paperback)
ISBN 978-981-98-1145-8 (ebook for institutions)
ISBN 978-981-98-1146-5 (ebook for individuals)

For any available supplementary material, please visit
https://www.worldscientific.com/worldscibooks/10.1142/14264#t=suppl

Dedication

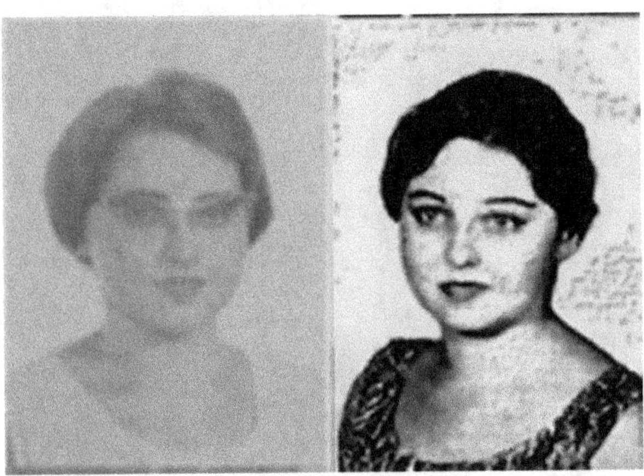

Figure 1. Dedicated to the memory of botanist Judith Georgia Colburn (~1945–~1975), born in Greensburg, Pennsylvania, a close friend who died young while employed in the Department of Botany, National Museum of Natural History (NMNH), Smithsonian Institution, a friend who knew me well enough from when we were both graduate students at the University of Oregon, to provide the main title for this book[2,3]. (With permission of the Sarah D. and L. Kirk McKay, Jr. Archives Center, Florida Southern College.)

Foreword

Oh Omnivore! The Path of an Eclectic Theoretical Biologist is a unique book, which introduces a reader to the scientific life journey of one of the leading contemporary theoretical biologists, Richard Gordon, who developed the most comprehensive and profound concept explaining biological morphogenesis and macroevolution. The whole story presented in this book allows us to follow the entire path of this prominent scientist and scholar. It is a very interesting and inspiring book for a wide range of readers spanning from the general public to the scientists involved in the development of modern concepts in biology and physics. In this book, Richard Gordon also shares his opinion about the practical aspects of scientific research such as the grant system, communications with colleagues, and publication policies. Written in an original form, with warmth and sincerity, and illustrated by many unique photos and images, the book will take a worthy place amongst biographical literature.

Richard Gordon is a prolific writer with over 200 peer-reviewed publications in a wide number of fields. During his scientific career, he performed broad interdisciplinary and cross-disciplinary work bridging biology with the fields of mathematics, engineering, physics and chemistry. In the current book, Richard Gordon gives an overview of different research areas in which he was involved in his scientific life journey. He goes through each decade starting with the 1960s and highlights the main points of his scientific career. The author goes through the milestones of his scientific life and engages the reader into his insightful explorations and discoveries. In particular, he invented the algorithm that was used for the prototype of magnetic resonance imaging (MRI), called initially zeugmatography. The broad educational background that included comprehensive knowledge of biology, physics, chemistry and mathematics, made it possible for Richard Gordon to be involved throughout his career in various theoretical and experimental projects that he describes with great inspiration and engagement. It resulted in the formation of the scientist of the universal kind defined in this book as an "omnivore", actually of the encyclopedically educated enlightenment-type scholar. The background that combines basic physical principles with the most recent advances in genetics, biochemistry, and biophysics became the basis of major breakthroughs in understanding origins of life, morphogenesis, and macroevolution processes.

The most exciting and far-reaching are the contributions of Richard Gordon to the development of theoretical biology, in particular, to the understanding of such fundamental phenomena as biological morphogenesis and macroevolution. In fact, embryogenesis, and generally biological morphogenesis, is one of the most scientifically demanding problems. The origin of the eukaryotic cell, followed by the emergence of complex multicellularity, and, as a result, of the complex developing organism, represent the most difficult and at the same time most inspiring challenges in biology. It rises to the most general problem of actualization formulated by Aristotle: to link the noumenal Platonic world to phenomenal reality. The processes of differentiation and morphogenesis are ontologically demanding in formulating them in modern scientific language and in developing a theory that would explain their principles.

Richard Gordon, by expanding and reformulating the idea of Conrad Waddington of an epigenetic landscape, introduced the fundamental concept of differentiation waves in 1985. These waves were demonstrated experimentally in collaboration with Natalie Gordon in 1989. They are associated with the morphogenetic organelle defined as the cell state splitter, the presence of which was demonstrated at the apical end of each epithelial cell in a developing embryo. The concept of differentiation waves goes beyond the molecular developmental approach that genes are the primary source of development. It introduces the breakthrough idea of the differentiation code that is based on fundamentally different principles than the genetic code and acts as an epigenetic principle of encoding morphogenetic structures. It represents a binary coding mechanism associated with mechanical differentiation waves that results in determination of cell type. This fundamental principle requires rigorous testing that may be realized only by the next generation of researchers.

The contribution of Richard Gordon to understanding the problems of origin of life and early evolution represents his other major theoretical achievement. The ideas of life before Earth, a transition to the eukaryotic cell, diatom cell periodicities and symmetries, and chromolinkers connecting eukaryotic chromosomes represent important milestones of his unique research career. Starting from the paper, where he invented diatom nanotechnology, he worked in such different fields as adaptive image processing, algal biofuels, computed tomography, neural tube defects, and social ethics.

Two major books of Richard Gordon — *The Hierarchical Genome and Differentiation Waves: Novel Unification of Development, Genetics and Evolution* (1999) and *Embryogenesis Explained* (co-authored with Natalie Gordon, 2016), both published by World Scientific Press, are the most challenging and comprehensive achievements in the understanding of biological morphogenesis. In the current book, the reader will find the inspiring stories of meeting with prominent scholars of the earlier generation including Isaac Asimov, Nicholas Rashevsky's biomathematics group, Robert Rosen, Conrad Waddington and many others. There are insightful examples of collaboration of the author with theoretical biologists of his and of younger generations including Lev Beloussov, Jack Tuszynski, George Mikhailovsky, Alexei Sharov, Bradly Alicea and many others. Organization by Richard Gordon of online seminars devoted to the problems of morphogenesis and evolution brought together many interested researchers and theorists and contributed to education of younger scientists. The book is highly recommended for a broad audience including not only embryologists and evolutionary biologists but also biologists of different profiles, historians of science, philosophers of science, researchers in various fields of science, teachers and students.

Abir (Andrei) U. Igamberdiev
Memorial University of Newfoundland
Editor-in-Chief, *BioSystems*/Elsevier

Prologue

I once owned and read *Pasteur — The History of a Mind*[5] by his lifelong lab assistant, Emile Duclaux (Fig. 2). I've had no such tag-along, so now that I'm 80, here I'll just have to do something similar myself, including my foibles. Pasteur had his, such as when he tried to improve French beer, but hated the taste of beer, a nonstarter. I don't drink beer either.

My life has been full of the common ups and downs, which I generally won't dwell upon, sticking mostly to the trivial and sometimes profound topics I've tackled. I once attended a talk by Isaac Asimov[6] (Figure 3) in New York City about 1969 on the need for interdisciplinarians, and left thinking "But who would

Figure 2.   Emile Duclaux (1840–1904) in 1890[4]. (Public domain Creative Commons Public Domain Mark 1.0.)

Figure 3. Isaac Asimov (1920–1992) in 1959[6]. (Public domain.)

hire them?" (cf.[7]). I suppose that it is for the few who might take this rocky road that I'm writing this, would-be theoretical biologists/"Renaissance men"/"Renaissance women" wanting to survive while doing their own thing. I've shown it is possible, if difficult. I've usually not lasted long in any one department at a university, racking up 12 departments during my 34-year career at the University of Manitoba, for instance. The only constant was the Department of Radiology, with most of the others being adjunct positions. After my tenure hearing for the Department of Pathology I was told that I could have tenure in "any department that would have me, except Pathology". The mathematically oriented Head of Pathology, William Thurlbeck (1928–1997)[8], whom I got along with, had left. As I had given a seminar on diatoms in the Department of Botany, I was known to them, and they were delighted in having their budget increased by one position. I survived in Radiology, even after I was hired by Botany, because I wouldn't accept the transfer otherwise. I otherwise would have lost access to medical grant funding. Thus, for a while I became probably the worlds' only Professor of Botany and Radiology. Finding a home as a theoretical biologist is difficult[9]. Few recognize it[10]. I became a full Professor of Botany and Radiology, "long overdue" according to Michael Sumner (Botany), in 1990.

Personal note: Finding an appropriate personal home can occasionally be awkward as well. Natalie is my third wife. It was Natalie's first attempt to make Montreal smoked meat: successful, but our meat slicer didn't cut it thin enough, so my teeth couldn't cut through it. I must have inhaled that piece from a sandwich, and immediately realized I was in trouble. In Natalie's earlier life, she taught Olympic swimmers, was a swimming pool lifeguard, taught police and learned a lot of medicine, including the Heimlich maneuver, which despite teaching it to over 1000 people, had never used it before herself in a real emergency. To our dog's delight, I explosively expelled that piece of meat. At 80, I then decided to go along with selecting our grave sites, right next to Saskatchewan berry bushes in an isolated, rural cemetery, joking how they will then be inaccessible, if delicious. She got me out of the hospital 4 days after a stroke, two days before lockdown, when COVID struck in 2020 and the Canadian "health" system became unsafe, with its push for Remdesivir, ventilators and MAID (Medical Assistance In Dying) and coercive mandates and lockdowns. Our daughter also has a degree in kinesiology, so with home computer therapy equipment, I recovered fully from the stroke in 6 months, having started from full paralysis. Natalie, a PhD epidemiologist, had read the Pfizer justifications for their "vax" and immediately spotted their incompetence, thus protecting us from the consequences, except from the wrath of some neighbors, one of whom claimed that our presence worsened his son's epilepsy and threatened to burn our house down. Natalie and I have been together now for more than three decades and thus our personal home suits us very well as long the world leaves us alone.

As a graduate student, I started organizing references to scientific papers I read on punched cards, and eventually discovered a series of computer programs that proved much better, from Professional Bibliographic System, ending up with EndNote, produced by Clarivate, who bought up most of the previous similar efforts. (Clarivate also offers the reference database Web of Science (WOS in the reference list)).Thus, along with tens of thousands of scientific papers in electronic form (sure beats those many filing cabinets I accumulated), I've managed to keep a reasonable record of my own.

So, I'll list my papers in chronological order, plus many of their spinoffs. There is much intertwining of dates, as I often came back to a subject, hopefully gathering new insights or applications. I've dropped hints here and there of yet unsolved problems I've encountered, some of which the reader might want to take up. Some of the images are crude, as they come from a time before we got used to computer graphics. Where available, I've included photos of the scientists and other people in my life who most influenced me, often relying on public sources, such as Wikipedia, for which no reader should encounter a paywall. I've tried to select photos from the time I knew them. (Where someone is deceased, I have tried to find their birth/death years.) (UC = unlocatable copyright, or no response from the copyright holder.)

Preface

I decided to be a scientist at age 10, reading many books from a local public library in north Chicago. My parents gave me the book *Mr. Wizard's Science Secrets*[13] when I was hospitalized for a week, perhaps from boils acquired at a YMCA camp, where I was led around by a horse, overheard the solution to the incantation "Owa tagoo siam", and correctly estimated the number of something in a jar, so that my "house" all got popsicles. I read *Mr. Wizard's Science Secrets* cover to cover, but then didn't do any of the experiments, perhaps because I already knew the results. I guess that's when I became a theoretician. I now call myself a "theoretical biologist", and all of my results are at best "best guesses". The only frustration is thinking ahead of others' problems, and trying to find someone to do experiments one proposes. I've done just enough experiments to know I can do them, but generally don't have the patience. My early hero was Roy Chapman Andrews (Figures 4, 6), author of *All About Dinosaurs*[14]. I was quite aware of the death of Albert Einstein (1879-1955) which occurred when I was 12, and picked up the mythology from my father, Jack Gordon (Figure 8, USA national doubles handball champion with his protégé John Sloan[15]), that only 10 men in the world understood him. As an elementary school kid, I collected butterflies, mounting them in cigar boxes (obtained, while my father played handball: boring to me), looked at snowflakes outside with a cooled microscope, raised stinky bacteria, owned Lionel toy trains and an Erector set, and fed my tropical

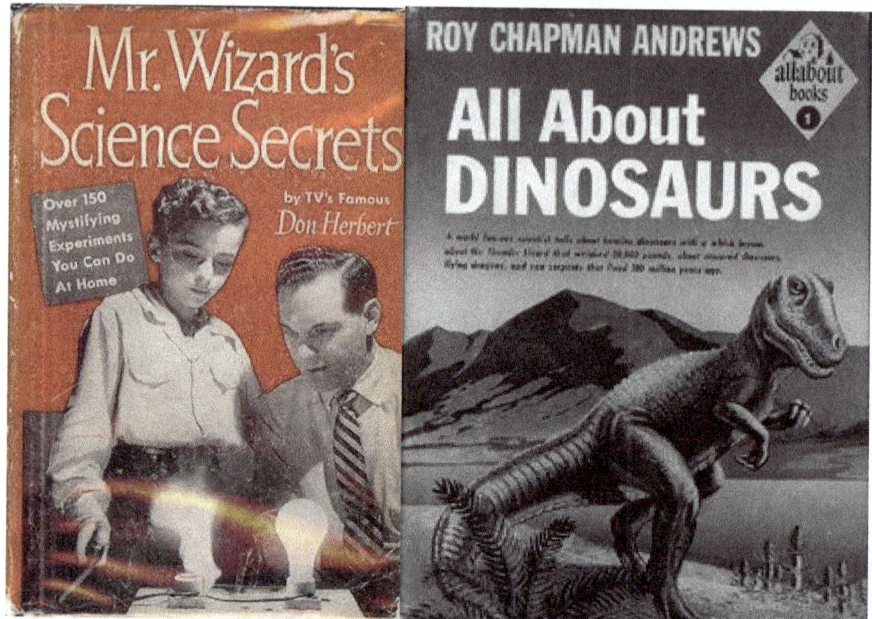

Figure 4. Don Herbert (1917–2007)[11] *Mr. Wizard's Science Secrets* (UC) and *All About Dinosaurs*[12]. (All About Dinosaurs with permission of PenguinRandom House.)

fish, some of which I bred, with *Daphnia* netted from a local construction pond. Guppies that I put in a local wooded pond disappeared, an ecology experimental failure. As an adult, I often had a pond, whether in an old bathtub sunk in the ground, or up to a $10' \times 10'$ dug one. As a kid, duckweed was a favorite, with *Vorticella* attached to its roots. Despite my nonexperimental bent, I checked the ideal gas law experimentally while in high school, to make sure it worked, using a bubble in a flexible tube from my aquariums and the $15 microscope my parents got me. I tried to generalize the ellipse by pulling a taut string around multiple nails (two suffice for an ellipse), but then realized that each arc was part of a different ellipse. I tried to generalize imaginary numbers, but found they were closed under my attempts. Thus, I learned enough math to be a mathematics spectator, but not participant. Most of Martin Gardner's puzzles in *Scientific American* evaded me. I was not a math genius, despite mastering long multiplication at a young age and being praised for it. (I recognized real mathematical genius when I read[16]. While mathematician G.H. Hardy, recognized this, he didn't invite Srinivasa Ramanujan's wife to England to take care of him, which was probably why he died young, an essential fact not mentioned in[17]. My wife, Natalie, has undoubtedly seen me past a few life-threating episodes.) When I played Scrabble, usually with adults, I insisted on cheating by using a dictionary, on the valid excuse that I then learned new words. As a teen, I usually carried a Lilliput Webster dictionary (Fig. 5) with me, to look up every word I didn't understand, especially on the hour long elevated/bus rides commuting across the city to University of Chicago/ Laboratory School classes. The painting of me was typical, not posed (Figure 11). I built a light box, whose ground glass was too thin when I picked it up by a corner, shattering it from its own weight, which might have interested me in cracking patterns and "turtle foam" (known to others as Voronoi diagrams).

Figure 5. A Lilliput Webster dictionary. (Etsy, cobweb, UC.)

I saw the green flash of the sun projected via my 3″ refracting telescope onto paper, and later, as an undergraduate, sketched sunspots as the Treasurer of the University of Chicago Astronomy Club often using their "antique 6.25‴" refractor from the 1890s"[18], of which there were then only two members: me and its President, John Musgrave[19] (Fig. 9), who got me interested in the history of science and kept my mind open and open to irony[20]: for instance, he deemed me Treasurer. (There were no other members and $0 budget.) John slept in the space between the roof and ceiling adjacent to the astronomy club room of the physics building. When I applied as a University of Chicago Laboratory high school student, I was asked to memorize the constellations, which I never could. Bucking such memorization tasks ruined me for organic chemistry,

Figure 6.   Roy Chapman Andrews (1884–1960) about 1920[21]. (Public domain.)

biochemistry, and taxonomy, but fortunately not for astronomy. I started grinding a mirror for a 6″ reflecting telescope, but never finished it, but did learn a bit of optics theory.

I tried to measure the height of meteorites using two cameras, with my brother Dan across Chicago, using a crude strobe I built with a motor twirling pie shaped sectors over the camera, to allow velocity measurments, but the film broke due to the cold night. I didn't try again.

As an adult, I've had two memorable deep experiences: a lightning flash a short distance away while I was peeing, which was so pure white and bright that all contrast in the washroom disappeared, and when I was doing the same beside a dugout in our quarter section in Alonsa R.M., Canada, a curious young crane hovered in my face. I was given a print of a crane by Canadian naturalist artist Robert Bateman (Fig. 7) as a result of the latter, as

Figure 7. Robert Bateman[22]. (Under the Creative Commons Attribution 2.0 Generic license.)

Figure 8.   Jack Gordon (1921–1994) and his sister Alice during WWII.

Figure 9.   John Brent Musgrave (1943–2015) at Great Slave Lake, Canada. (By the author.)

a present. Although I was never good at it, an appreciation of art permeates my story.

Librarians, whom I respect, perhaps due to a childhood aspiration to establish a neighborhood library of brochures, in retirement remain my major contact with academia. One librarian I remember, head of the Children's Library in the downtown Chicago Public Library, Bernice Kirkpatrick, got me access to the adult books, where I admired and tried to understand an English version of Newton's *Principia*[23], well before I understood calculus. My "job" there was to watch my younger sister Helen and brother Dan while our mother, Diana Gordon (Figs. 10, 11), attended classes at the nearby Art Institute of Chicago, at which I also learned to paint and sketch at kids' classes. Bernice visited us at home, and the overall experience engrained my blindness to a person's color or race. As an adult I've felt comfortable in many cultures. Another important person, who's nearby home I frequented as a teen, was *Life* magazine photographer Wallace Kirkland (1891–1979)[24]. As is typical of photographers, I couldn't

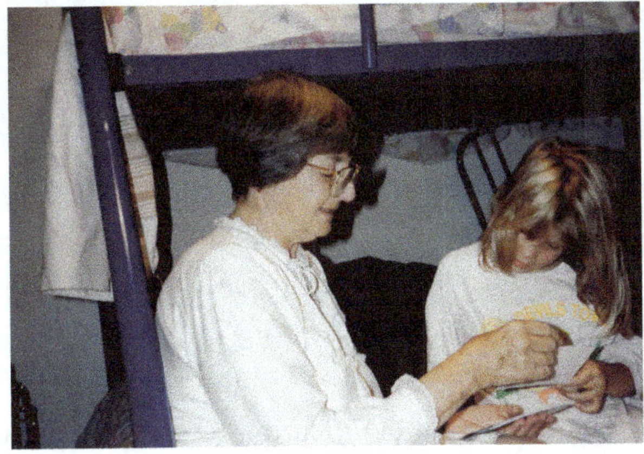

Figure 10. Diana Gordon (1922–2018) with granddaughter Lana. (With Lana's permission.)

Figure 11.   An oil painting of me by Diana Gordon done when I was 20 in 1963.

find any representative photo of him himself, while lots of his photographs are online. We shared in spirit *The Lure of the Pond*[25].

An image of the phases of Venus like Figure 12 has stuck in my mind since my freshman year in high school at John Dewey's University of Chicago Laboratory School[27] (Fig. 13) which was well ahead of the time as an "integrated" school, where I dated a very black Carol Williams who unfortunately was in love with an on duty marine. My father reacted by merely pointing out the difficulties with mixed marriages. I was given an opportunity to pass out of first year high school general science, but failed the exam because I presumed that all phases of Venus were the same size. As I had a Unitron telescope[28] as a kid (and later designed one[29]), I would have changed its magnification to fill the field of view, so in retrospect I was correct, but nevertheless had to suffer the same material repeated *ad nauseum* every year in elementary school. That "mistake" cost me a year of boredom, for which I chewed out

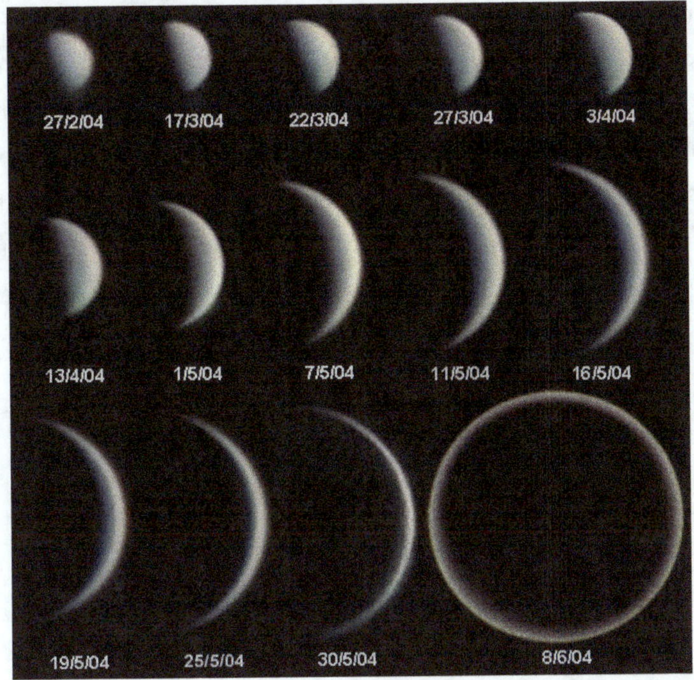

Figure 12.   Phases of Venus[26] .(Public domain.)

Figure 13.   University of Chicago Laboratory School[27]. (Public domain.)

the school in requested year-end comments, most other kids complaining about the lunch food. I think they dropped general science for high school students as a result. In shop, I made a conical brine shrimp hatcher for my tropical fish, and a telescopic pinhole camera, which worked. In home economics I baked a flat cake, substituting honey for sugar. I spent summers reading the next math textbook and "passing out", or taking a summer math course. One summer, every lunch was from the hot dog stand of an old Polish man. I left for the University of Chicago after 3 years, never graduating high school. I had been voted President of the high school German student club before I left, but then was gone. The only problem with being a young college student was that I was too young for the girls. Perhaps I didn't do my share of washing the perpetual sink of dirty dishes in an all-male shared apartment in Hyde Park.

In college, I kept switching my undergraduate major between chemistry and physics, and then realized that I could graduate in 3 years with a math degree, as I had promised myself to take one math course every semester. I was never a darling of the math professors. Graduation was uncertain to the last minute, because I got a D in a paleontology course (Andrews' influence notwithstanding, Fig. 6), which I especially enjoyed on the field trips, except for the later boredom of classifying brachiopod fossils. They let me graduate anyway.

I thus ended college with a hands-on paleontology course. Highlights of that course were a field trip collecting fossils (because of my age, the older students deferred to me by avoiding bars, which I could not enter), and an anecdote about a young fellow scheduled to give his debut lecture right after a film was shown of rhinoceroses mating. Years later I found myself in a similar situation, giving my talk on AIDS prevention at a gay conference in Winnipeg right after a talk on how a glass coffee table allowed more certain safe sex than condoms[30]. The latter was my only paper with a caveat:

> *"Editor note: The following article on a topic of considerable individual and public health significance presents disturbing information and leads to conclusions of a controversial nature. Letters to the Editor with comments and discussion on this important topic are welcome."*

No discussion occurred.

Thus, I started graduate school at age 19, and was chagrined that it took me 4 years to complete the PhD, not 3, after 3 for high school and 3 for college. Nowadays, with graduate students retained for cheap labor, the average duration is about double, at 6 years[31]. I never felt exploited as a graduate student.

Contents

Chapter 7   2020s   **215**

Figures

Chapter 1

1960s

1. Contaminated Meteorite (1964)[32]

I worked in the lab of E. Peter Geiduschek[35] (Figure 15) as an undergraduate for a summer, running D.N.A. "melting" curves[36], and learning about ultracentrifuges and electron microscopy, in the nascent field of molecular biology, then in Biophysics at the University of Chicago. Yes, DNA was still abbreviated that way. An impressive lecturer from Argentina, an electron microscopist, discouraged me from joining him as a grad student, because of the politics there. In those days, we brought our lunch in paper bags,

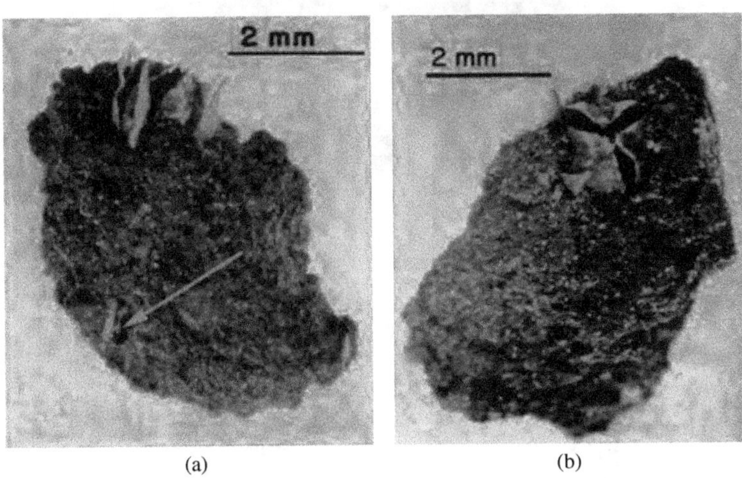

Figure 14. (a) Seed capsule embedded in Orgueil meteorite fragment. Arrow points to protruding stem. (b) The same fragment, end view"[32] with permission of the American Association for the Advancement of Science. Contrast enhanced with histogram equalization using ImageJ[33].

Figure 15.   E. Peter Geiduschek (1928–2022)[34], with permission of Annual Reviews, copyright 2009. All rights reserved.

Figure 16.   Robert Rosen (1934–1998)[40], public domain.

which I was in the habit of popping. When people came running with fire extinguishers and I got chewed out, I never did it again. Much later, when I was a Professor, I brought forgotten lunch to one of my kids in nursery school at the University of Manitoba, to everybody's much amusement, carrying a Sesame Street lunch box. While at the University of Chicago, I attended lectures in Mathematical Biophysics organized by Nicolas Rashevsky[37] and read his book[38]. I later postdoced with his student Robert Rosen[39] (Figure 16).

Figure 17.   Helmut Hirsch[41], UC.

I got into Biophysics when my long-term best high school friend Helmut Hirsch (Figure 17) and I snuck into a University of Chicago open house for high school students at the Enrico Fermi Institute, across the street from where the first atomic reactor was built. I was absolutely fascinated by Bob Uretz's microbeam work in which a cell would eject a part of the chromosome that he had just irradiated with UV light[42,43] (Figure 18). I lingered. The fascination turned me into a biologist, even though of the only biology course I took at the University of Chicago I only recall distasteful frog dissection. The head of the Department of Biophysics, Ray Zirkle (Figure 18), started this work[44], and I got the summer job with E. Peter Geiduscheck (Figures 15, 18). Over the years I kept abreast of microbeam work, and applied it to the colonial diatom *Bacillaria*[45]. Helmut's work started out in cat vision.

This paper[32], on a contaminated meteorite was my debut into the scientific literature, if only as a footnote, of which I was proud. I was an undergraduate in Edward Anders'[47,48] (Figure 19) lab at the University of Chicago, who at that time offered an undergraduate course on quantitative chemistry that I took the summer before starting college. I was then hired to prepare samples to be

Figure 18.   "UChicago biophysics faculty, 1964 [a year after I left for graduate school]. Top: Bob Haselkorn, Bob Uretz [(1924–2018)], Ed Taylor, Peter Geiduschek, Bob Haynes; Bottom: William Bloom, Ray Zirkle and John Platt, a physics professor." From[42], UC. Edwin W. Taylor's apparently unpublished work on amoeba whose cytoplasm in pseudopodia still moved properly when the membrane was removed impressed me. I visited Albert Crewe when he was imaging single uranium atoms[46], and attended with many others a lecture by James Watson showing much lower resolution "writing" at which he put his audience down with "I suppose that you came here to see a Nobel Prize winner".

Figure 19.   Edward Anders, 1965[49], with permission.

analyzed by students in quantitative chemistry courses, a demanding task, as they couldn't do any better than I did. I learned how to be experimentally precise, getting 4 decimal points out of an ancient mechanical balance, often sleeping overnight in my preparation lab over a large teaching hall, quite some facility for a freshman, and identifying some crystals Helmut brought me by color, when he took introductory chemistry, which I skipped. Afterwards, I was chewed out by the head of Chemistry, for not reporting my hours, while I got the job done in less than the time allotted, which seemed fair to me. He made an expensive mistake in accusing me, as the University of Chicago never got an alumnus contribution from me as a result. Anders also gave me my own desk in his lab.

After a brief conversation in an elevator in the Enrico Fermi Institute at the University of Chicago, Anders (Figure 19) invited me to work in his lab on organic matter in meteorites. I designed and had built a vacuum apparatus for sublimating and precipitating sublimate from small specimens, after having spent much time in the glass blowing shop. It imploded, destroying itself, but Anders got the bug. He received a sample of the Orgueil meteorite, which fell in Orgueil, France in 1864, and hired a Japanese postdoc familiar with the then new HPLC (High Pressure Liquid Chromatography) to do things right[50–55]. It was now 1964, one century for the meteorite sitting on the shelf in an Orgueil museum. Anders checked a black substance embedded in a sample which proved by x-ray diffraction to be coal, and a plant bud local to that area (Figure 14). I showed that the meteorite was held together by animal glue[56]. He pointed out in untampered samples that the other amino acid content was no greater than a fingerprint. Thus, it was a fraud, which the fraudster didn't live to see discovered. I learned honesty in science from Ed Anders, and have been careful and honest to myself since then.

Sometimes my insistence on careful and accurate work, especially in multiple rewrites of papers, has driven people in my lab nuts. I found a microscopic "panspermia" swimming in an extract from the Orguiel meteorite, presuming it was there due to my not using sterile techniques and the meteorite providing nutrients. I should have published this result rather than being just amused, as it was decades later rediscovered[57–59]. I have written about Ed Anders in the Preface to *Habitability of the Universe Before Earth*[60]. I also jammed an unjammable Marchant mechanical calculator[61] that a frustrated salesman was demonstrating to Anders. Anders gave me a piece of Gunflint chert[62], which eventually inspired an indirect measurement of ancient atmospheric N_2 using cyanobacteria[63–66], originally formulated with astronomer James Walker (Figure 20) on a visit to the Arecibo Radio Telescope in Puerto Rica (Figure 21) before it collapsed, during which I got to travel by gondola to walk the arced receiving antenna 400' above the radio mirror. I was told that underneath the reflector, plants grew with as much light as available on Mars.

Follow-up paper:

Life has been "discovered" in meteorites at least half a dozen times over the centuries. Thus, the title: *Recurrent dreams of life*

Figure 20.   James Walker (1939–2022)[67] with permission of Donna Wessel Walker.

Figure 21. Arecibo Radio Telescope (1963–2020)[68] Creative Commons Attribution-Share Alike 4.0 International license.

in meteorites[69]. (Cf.[70].) Jesse McNichol, now a faculty member in oceanography (Figure 22), and I also wrote *Are we from outer space? A critical review of the panspermia hypothesis*[71].

I am concerned that early returns of material from asteroids suggest that amino acids in them are racemic mixtures, whereas most of the carbonaceous chondrite meteorites, thought to be samples of such asteroids, show enantiomeric excesses, biased towards L-amino acids predominantly found in Earth's

Figure 22.   Jesse McNichol[72], with his permission.

organisms. This apparent contradiction is mentioned in *Origin of Life via Archaea: Shaped Droplets to Archaea First, With a Compendium of Archaea Micrographs*[73]. Something may have caused the enantiomeric excesses after meteors left their asteroids[74], perhaps the polarized light bouncing off bodies of water, or perhaps they are Earth-contaminated. Even table salt has been found on a meteorite that should have been pristine[75].

During the 1997 Red River Flood[76], I recall sandbagging at midnight in Winnipeg, with the river 2 feet below on one side, submerged trees blocking chunks of ice, and 20 feet of sandbags protecting the other side. Overhead, as I placed sandbags, was a bright comet that I noted contained even more water[77]. My kids had moved my 2000 books from the basement to the second floor of our home, when they weren't filling sandbags or helping others move their furniture. One soldier remarked in French in Natalie's presence at the pleasant swaying of girls' breasts as they passed sandbags in a line, and blushed when Natalie replied in French. We helped the people running a food truck for volunteers to provide kosher food donated by local stores for Jews celebrating Passover.

People with heavy equipment built a 15 mile dike in 3 days[78]. One amusing note was because we both worked at the hospital, emergency preparedness officials had both Natalie and I list our potential useful abilities in case the city was inundated. Since Natalie was informed as an experienced mother and La Leche League member (a support group for nursing mothers), she was assigned to report to the low-risk maternity ward to free up nurses by helping new mothers. I was told to take my kids and get out of the city.

Natalie hoped to teach in a new school that was planned for teaching midwives, then cancelled by an incoming government. We later sent a shelf of midwifery books to Afghanistan for the midwifery association there, via Books With Wings.

2. On Stochastic Growth and Form (1966)[79]

At that time, the hidden quotas on Jewish students hadn't faded away. So, after getting some advice, I started graduate school at the

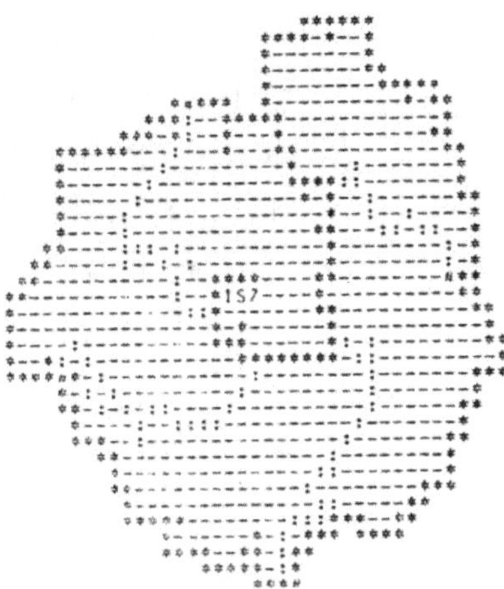

Figure 23. "The best spiral obtained under local growth"[79]. The colons represent consecutive growth rings. For indefinite spiralling, I had to add a long-range (global) interaction with the growing cell nearest the outer shell (*). Local/global interactions prevailed in my later models of morphogenesis (Figure 24). Copyright held by author.

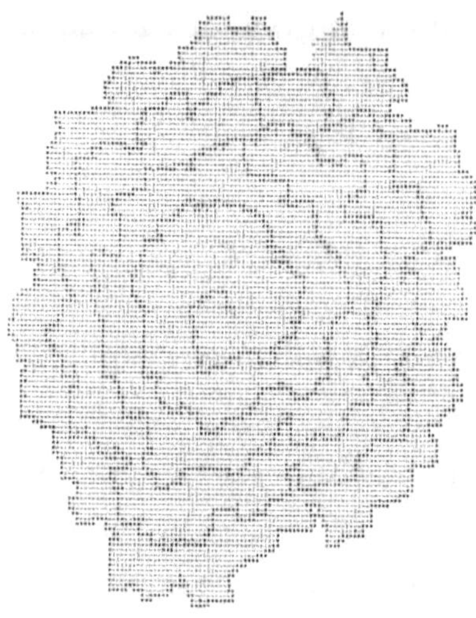

Figure 24.   A global/local simulation of 2D "snail"[79]. The growing tip (*) stochastically varied the number of uncovered dashed (-) cells it was adjacent to. Copyright held by author.

University of Oregon with Aaron Novick[80] (Figure 25) at age 19, despite invitations from Stanford (unimpressed at a visit) and Yale (no girls). Aaron was formerly at the University of Chicago. I decided to start with a light load, and in the middle of my first term was drafted to serve in the Vietnam War, as I was not exempt, being a part-time student. Aaron got me registered for a reading course in the middle of the semester, and I didn't become one of the 50,000 Americans killed in Vietnam. I was aware, through my brother Dan, who was then a radio broadcaster, that Ho Chi Min praised the American revolution, perhaps contradicting the news[81], an observation he did not censor, despite US Government encouragements to do so. Throughout the war, I marched against the Vietnam War with other students, presuming I accumulated a large CIA file. I tried to keep peace between the protestors and the kids in the Reserve Officers' Training Corps (ROTC).

Figure 25.   Aaron Novick (1919–2000) in Oregon's Cascades. (Photo by Gary Tepfer), with permission of Frank Stahl[82].

I recall joining a small group being spoken to at the University of Oregon by a US Government representative, and found his attitudes distinctly un-American. The University of Chicago had instilled a healthy respect for the accomplishments of the founding fathers of the USA. I delved into parapsychology until a speaker on it emphasized experiments done on mentally influencing the growth of beans. On afterwards looking up the original papers, I found them incompetent.

Aaron Novick had worked with Leo Szilard[83] on developing the chemostat[84], which I much later used to inspire the compustat, a device that might allow deliberate alteration (higher level "directed evolution"[85]) of the shapes of diatoms[86]. Given the new classification of diatom shapes in terms of Legendre polynomials[87], a compustat could be given instructions to select changes in shape

Figure 26.   Terrell L. Hill (1917–2014)[92], with permission of SpringerNature.com

to another Legendre class of shapes, or fail to accomplish such directed evolution, perhaps improving our poor understanding of shapes of single-celled organisms and the limits of their malleability. No one has yet built a compustat.

I then dropped out of The Institute for Molecular Biology at the University of Oregon, switching from Aaron Novick (Leo Szilard whose death that year caused him much stress) to Terrell Hill[88] (Figure 26) in Theoretical Chemistry, disturbed that in a lab full of people studying the bacterium *E. coli*, I was the only person who checked with a microscope that they were really there. Also, I never could plate *E. coli* properly.

Partly inspired by a high school essay on nuclear war, reading *Hiroshma*[89], Szilard's attempts at civilian control of nuclear energy[90], seeing Harry Belafonte speak at SANE[91], and meeting refusniks in Moscow, I developed a bit of a pacifist streak.

At the time, Terrell had 3 graduate students: George White, Joel Keizer, and me, who worked independently and got along

Figure 27. Joel Keizer (1942–1999)[93], UC.

well. Joel (Figure 27) also started in molecular biology. He invited me to give a seminar at University of California Davis after he graduated, and I went to the conference (Nonlinear Dynamics in Biology and Chemistry), in his honor, after his early death, driven there from San Francisco by George, but was out of place as the only one except his widow from his previous life in Oregon.

I think it was my friend, Susan Meschel[94] (Figure 28), then a graduate student in Quantitative Chemistry at the University of Chicago, who sent me a journal issue containing[95]. I spent a summer in Berkeley, California, taking UCBerkeley courses on neurophysiology, probability theory, and rat behavior, getting A's in the former two and an F in the latter. While there I got to meet Magorah Maruyama[96] (Figure 29), author of[95], his French wife, and their new baby. Terrell had just written books on small systems thermodynamics, in which stochastics plays a large role[97,98] (expanded by[99]). Maruyama had included a crude, deterministic model for morphogenesis in his paper, with interacting "cells". As embryos are small systems, and cells even smaller, I figured that adding a stochastic element to the behavior of the embryo's cells could test for the robustness of such models. I bought D'Arcy Thompson's[100] 2nd edition of *On Growth*

Figure 28.   Susan Meschel in 2023, with her permission.

Figure 29.   Magorah Maruyama (1929–2018)[107], UC.

and Form[101] (Figure 31) at a Berkeley used book store, and was hooked, later gleefully noting all the spring tree buds that would morph into leaves. I came up with a stochastic model of a 2-dimensional "snail", and tried to simulate it on graph paper tossing coins. It was tediously slow. So, I decided to learn how to use/abuse a computer, phoning someone to learn where the on-switch was on a card punch my first day, on a weekend, for an IBM 1620, which could handle about 300 "cells" per second, faster than me, but orders of magnitude slower than today's computers. I eventually produced a boxful of paper output on a later computer, an IBM 320, which my friend Judith Colburn (Figure 1) and I took to a photographer to copy page by page onto movie film, 2500 frames as I recall. On the last frames I sketched a fish eating the "snail". As it wasn't chemistry, I did this all behind Terrell's back, but made up a project that resulted in my PhD thesis and a few papers, to justify to him my use of a computer[102,103]. To my surprise, Terrell was delighted when I confessed. He had just been elected to the US National Academy of Sciences, so my first paper *On stochastic growth and form* was in PNAS[79] (Figures 23, 24), sponsored by him. Quite a boost for a nascent scientist. I think that Max Braverman reviewed it. He was an ex-Marine, who had a trailer near his work at a Pittsburgh hospital as a medical physicist, in which he had tanks of marine hydroid colonies to study their morphogenesis and worked with a fellow who did computer simulations of their patterns at Los Alamos National Lab[104–106] (Figure 30). I was his home guest. Max tried to teach me to drive a car in the Pittsburgh mountains where he lived, while he

Figure 30.   Max Braverman (1929–2004)[108], with permission of Robert J. Baran.

Figure 31.   D'Arcy Thompson (1860–1948), about 1890[100], public domain.

was drunk. I refused and we both lived. Max and I concluded that simulating morphogenesis was not so hard, but getting the mechanisms cells actually use was difficult.

By the way, in both classes and at invited seminars, I developed the bad habit of conspicuously falling asleep and then annoyingly waking at the end with relevant questions. My relationship with Terrell somehow survived this and my disinterest in his tennis. Reading outside the chemistry building under an oak tree, I was pelted with acorns by a blue jay. I responded by throwing a handful of them up at the bird.

By this time computer centers were running faster computers, with paper output rather than punched card output. One day I decided to plot a framed graph using characters, and made a programming mistake, so that the top border line on wide, continuous form paper, mostly dashes, was printed repeatedly at the top of the page on a hammer printer[109]. Some clever engineer had decided that if the paper wasn't moving, then the ribbon didn't have to move. I thus cut the page-wide long, cylindrical $90 ink ribbon all the way across and was banned from the computer room since I could do so much damage from the outside. It reminds me of Wolfgang Pauli, a theoretical physicist who was renowned for how much equipment broke when he visited a lab[110].

In any case, this snail paper launched my career in morphogenesis. My first box of printed "snails" was made with a bug in my computer program, whose effects I could not determine. So, although the output looked right, I threw out the whole box, fixed the bug, and did it right, thanks to the self-honesty that Ed Anders (Figure 19) had taught me.

I started with Fortran 2, graduated to Fortran 4, later taught Pascal to medical students at the University of Manitoba, all now dead computer languages. I made the mistake of deferring to my students for programming, and was thus left behind in my computer skills. Now retired, I struggle with Mathematica, the days of meeting at a computer center long over, where one could get help from other students, on any question no matter how trivial, without charge (typical turnaround time from batch processing computers was 1 hour). I am hoping to learn Python, so that I am not so dependent on youngsters in my old age.

I thus once "spoke" Fortran, which actually counted for the farcical graduate student foreign language requirement. For instance, I passed Russian by having to translate a paragraph of chemistry after 3 weeks of studying the language. It was full of cognates and a dictionary was allowed. Needless to say, Russian wore off well before my later trips to the USSR and Russia and my collaborations on papers with many Russians, who live either in Russia or North America. My grandparents were from Russia and Poland, but language is unfortunately not transmitted via our DNA genomes, yet[111].

3. Steady-State Properties of Ising Lattice Membranes (1968)[103]

As I said, I felt that I had to make up a proper statistical mechanics problem to justify my foray into morphogenetics in a chemistry department (although I later learned that Richard Noyes was working on the Belousov-Zhabotinskii reaction[114] in Chemistry, which Boris Belousov[115] could only get published in an obscure journal[116], an early lesson on how major breakthroughs in science were suppressed if counterintuitive). At that time it was well known that at equilibrium a 2D or 3D Ising lattice underwent a proper phase transition, and forays into steady state statistical mechanics were being attempted[117]}. (Ironically, Ising[118,119] devised his model in 1D for phase transitions, which proved not to work.) But, the phase transition was mathematically difficult. Terrell (Figure 26) was working on it at the time[113,120], so I decided to test it precisely by Monte Carlo simulation[121], which became part of my PhD thesis. It also provided a possible escape route to Sweden, were I drafted again[122], proposed by Terrell. I read a book I can no longer locate: *The Light in Sweden*. Terrell and I had shown that statistical mechanics did not work for discrete steady state systems

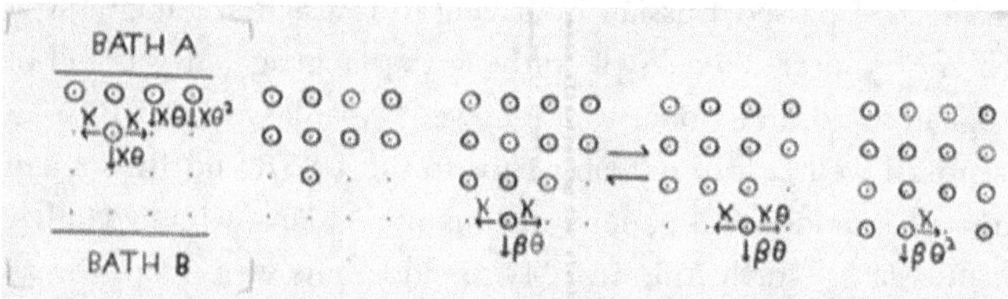

Figure 32. The Ising lattice as a membrane between Bath A and Bath B, from my PhD thesis[112]. Similar events, jumping to Bath A or back (not shown), would occur with this sampling of states. See also[113].

(as implicated by Edwin T. Jaynes[123]), so I proposed simulating an Ising lattice membrane. The other thesis part was the 2D snail. To my surprise, when I wrote up the Ising membrane work for publication, Terrell took his name off of it[103], giving me full credit. The result I got was curious: when bath concentrations allowed a phase transition to occur, the phase transition severely cut off flux across the Ising lattice membrane. This switch-like behavior was an important lesson in nonlinear behavior of some systems. I tried to apply it to the hydrogen in palladium system, but doubt that has stood the test of time. I bought 100 toy compasses, and put them in a square array to watch an Ising lattice in action.

I met Terrell's collaborator Ora Kedem[113,124] (Figure 33) later at a small meeting in Basel, Switzerland[125], who may have been instrumental in getting me to visit the Weizmann Institute, my first visit to Israel. Some of her work was on desalinization. My Swiss presentation on correlation for 3D electron microscopy tomography has been lost.

Figure 33.   Ora Kedem[126], Public Domain.

While I was a graduate student in the 1960s, I had to propose some research topics other than my thesis work. I suggested the dynamics of schooling fish. I recall thinking my career as a scientist was over, when the committee members came out and shook my hand. Following this schooling fish infatuation, I took the long, multiple train ride from Eugene, Oregon to Tampa, Florida to visit Charles Breder at the Mote Marine Laboratory, who had retired as Head of the New York Aquarium, before he had written an article on schooling fish[127]. There I observed schooling mullets off piers, small stingrays at the water's edge at dawn, Breder feeding local racoons, and got to ride the bow of the shark lady's (Eugenie Clark's) boat[128]. I did the first computer simulation of schooling fish, in the configuration of circular milling behavior, but computers were still too slow, except for me to note that the simulated school slowed down (cf.[129–132]). The output, printed in characters like / \ on paper only indicated the orientation of the "fish" in a plane, to the nearest 45°. Breder confirmed that real fish mills do slow down. This experience led, in combination with my interest in embryology and pigment cells (I was a charter member of a pigment cell society and thereby developed an awareness of melanoma), to keeping up for a while on research on generally pigmented larval fish. Unfortunately, Breder and I never published our thoughts, and what became known as "swarming behavior" was mastered two decades later on much faster computers[133], to the point where murmurations of thousands of birds in 4D can be simulated ("boids":[134,135]). I have seen huge murmurations of birds on our travels and observed many swarms of insects. I still have the long-term letter correspondence, handwritten, between Charlie and me. Boids may prove to be important for computed tomography imaging for detecting breast cancer[136].

4. Blinking Christmas Tree Lightbulbs in Series (1969)[137]

Science is about curiosity and problem solving. Every once in a while, I come across or formulate a problem that has no practical value, and tackle it just for fun. At the time this was written, blinking Christmas tree lightbulbs were thermal switches, with bimetallic strips like those in aquarium heaters, many of which I had handled and fixed when I collected and bred 6 tanks of tropical fish as a teenager. I wrote it up[137] for a Christmas edition of Jim Danielli's[139–141] (Figure 35) Center for Theoretical Biology annual report, north of the main campus of SUNY (State University of New York) Buffalo, with my experimental work on a string of lights done previously at Ted Puck's[141] (Figure 36) Department of Biophysics at the University of Colorado in Denver. While working in Denver, I commuted to Boulder, Colorado, where I lived and had the computer center, in a chauffeured car. I learned about the incredible ability of the mind to adjust to distorting or inverting glasses from a frequent fellow passenger who had volunteered to wear them.

Figure 34.   One blinking bulb[138], UC.

Figure 35.   Jim Danielli (1911–1984)[144], UC.

Figure 36.   Theodore Puck (1916–2005)[145], UC.

Danielli once gave a seminar on a remarkable self-observation: he wrote down the time he thought it was, at which he awoke at night, and then checked a clock: his error was at most about 20 minutes. He posed the problem of what mental clock or environmental cue allowed him to do this? This problem has been with me since the late 1960s. At 80, I have finally realized a possible solution: the regular filling of his bladder, a periodic, nonlinear, triggering event. If you are old enough, you can check your accuracy yourself.

Danielli had us join a press conference at which he announced the synthesis of life[142]. He asked us to announce our new ART algorithm[143] for what we then called "reconstruction from projections" at the same event. News of his work went around the world. Ours went nowhere.

In retrospect, blinking Christmas tree lightbulbs in series is an example of strong local/global interactions. When one light bulb heats up enough to turn itself off (the filament was connected to the bimetallic strip), power was cut to the rest of the light bulbs because they were in series circuit, each of which immediately turned off. The curious problem was: what is the frequency of on/off versus n bulbs compared to a single bulb? The time on for the string, empirically, went down exponentially with n.

I tried to solve the problem mathematically, but couldn't get past $n = 2$. Mathematica might help here.

Now that I've learned a bit about graph theory and mathematical networks, the problem can be generalized. Suppose that each node in a wired network has a blinking Christmas tree lightbulb. Let current flow between any two nodes of the network, with two nodes connected to a battery. What is the behavior of portions or the whole of the network? This involves both the on/off behavior of the individual nodes, but also resistances across the network, which determine rates of heating. I leave the problem for the reader with time on their hands. I toyed with similar iron wire models for nerve propagation in networks[146–148]. A chapter *Irritable Protoplasm: Forerunners to Differentiation Waves* in[149] reviewed the history of such models. My wife, Natalie (Figures 37, 38), has been a close collaborator in embryology.

Figure 37.   Me and Natalie K. Gordon, nee Björklund in 2018 at Gulf Specimen Marine Lab & Aquarium.

Figure 38.   Natalie's favorite picture of me, taken in Georgia, USA, while on a visit to diatom ecologist Kalina Manoylov[150] (Figure 39), a visit made many times.

Figure 39.   Kalina Manoylov, with her permission.

Other useless projects I worked on were a computer program that played off two strategies for the game of pentominoes[151], in a computer that only did batch processing. I tried to solve the question posed by Martin Gardner, that $n^2 = \Sigma_{i=1}^{24} i^2$ only for $n = 70$ (and trivially for $n = 1$). Can one put all 24 squares on a 70×70 board? I could prove that certain squares could not be at the edge of the 70×70 board. Later the problem was proved impossible by[152]. (I was not good at Martin Gardner's puzzles in *Scientific American*[153,154].) So, I leave the reader with a related problem: can it be done with periodic boundary conditions, which eliminate all edges? If so, could the arrangement of squares tile the plane?

I indulged in a bit of science fiction with Banne Vukotić (Figure 41), suggesting that someday we could make mini-Earths with black holes as their cores[155], though we do take prospects for panspermia seriously[156–160].

I tried to separate Fred Hoyle (Figure 42) from his colleagues on the question of panspermia[161]. I occasionally indulge in political satire[162] or commentary[163–168] (Figure 40). I broke some

Figure 40.   In Winnipeg, me in Afghan chitrali hat and Natalie in a burka counter-protesting, supporting Canadian troops in Afghanistan[170], UC.

Figure 41.   Banne Vukotić, with his permission.

Figure 42.   Fred Hoyle (1915–2001) statue[172], under the Creative Commons Attribution-ShareAlike 2.0 license.

bones in my foot going down the stairs to my basement lab, and wishing to get back to work, to get me out of the house Natalie negotiated with the Workers' Compensation Board for rental of a scooter, much cheaper for them than my monthly salary.

I learned the difficulties of triggering motorized doors while riding it with my crutches and trying to get through them before they closed. Published complaints in the university newspaper[169] were "rewarded" with an appointment to the University of Manitoba handicap committee and died there.

Fred Hoyle (Figure 42) has been a bit of a foil over the years, though he preceded me. For instance, I took up his junk-yard challenge to the origin of life[171], showing later how life could have started at chemical equilibrium[73], contrary to the common assumption of astrobiologists that it required an energy input.

I can contradict or correct myself. While Alexei Sharov and I showed that life might be twice as old as Earth[173], oceanographer George Mikhailovsky (Figure 43, with whom I'm presently working on the Earth period called *The Boring Billion*[174]) and I showed that due to the very short days Earth had just after the Moon was formed by collision with Theia[175] (but cf. [176]), as short as 2 to 6 hours, life starting on Earth was quite plausible[177,178], with 10^{11} days available during the origin of life, possibly as many PCR (Polymerase Chain Reaction) cycles. "Typically, PCR consists of a series of 20–40

Figure 43. George Mikhailovsky[185], under a Creative Commons Attribution 4.0 International License.

repeated temperature changes"[179], a bit fewer when done by short-lived humans. For instance, PCR-like reactions could generate many different molecules such as peptides, RNA and DNA with that many short days available during the origin of life[73]. This happens because there is a small chance of "error" (abiotic "mutagenesis") at each PCR step[180]. Natalie and I showed how membrane peptides could have thereby been abiotically selected[73,181].

My intuition that a hexagonal array of pixels is better for computed tomography than a square array proved wrong[182].

George and I are also estimating whether or not my postulated evolutionary advancement of "continuing differentiation"[183] might have occurred during the Boring Billion[184].

5. Polyribosome Dynamics at Steady State (1969)[186]

Ribosomes had been newly discovered, and many would attach to one messenger RNA (mRNA), the structure called a "polyribosome" or "polysome" (Figure 44). I had an interest in queuing theory, on which I sat in on a course at Columbia University, and traffic flow theory. I visited Luna B. Leopold[188], with whom I discussed kinematic waves[189], which occur in rivers and traffic flow. This got

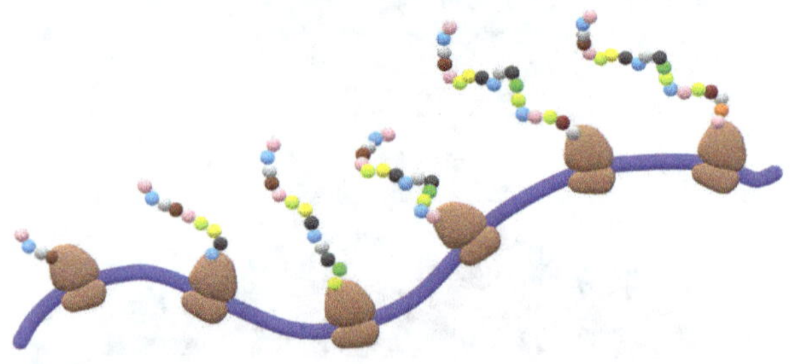

Figure 44.   Polyribosomes[187] under the Creative Commons Attribution-Share Alike 4.0 International license.

Figure 45.   Luna B. Leopold (1915–2006)[188], public domain.

me interested in geomorphology, and correspondence about meandering rivers with Einstein's son, Hans Albert (1904–1973)[190], and others on the branching of river deltas, presaging my later work on underwater saltation[191].

I was at the University of Colorado, which had the ability to make a film movie directly as computer output. So, I made movies of ribosomes jumping onto one end of an mRNA, across, and off the other end[186]. No ribosome could get past the one ahead of it. These were the days (and previously) when one hung around in a computer center waiting for your output. Some of the others were computer science students, then a new field. It became obvious to me that they had no idea what computers were good for. Now they seem to rule the world.

The functioning of polyribosomes is still an active area[192].

Chapter 2

1970s

6. ART and the Ribosome: A Preliminary Report on the Three-dimensional Structure of Individual Ribosomes Determined by an Algebraic Reconstruction Technique (1970)[193]

At that time the mechanism of the ribosome was a mystery. I worked with Robert Bender (Figure 48), a lifelong friend since then, trying to reconstruct ribosomes from 6 electron micrographs taken at different angles (Figure 46). He became a venture capitalist and bought me a painting by Winnipeg artist and friend Franz Visscher[194] (1929–2014). For unknown reasons the scientists who took the tilt series of electron micrographs dropped out, but left their micrographs with us. Trying to repeat this with a tilt stage, I gained a great respect for electron microscopists, and had met Humberto Fernández-Morán[195] (Figure 47) and Albert Crewe[196] (Figure 18) while I was an undergraduate at the University of Chicago. Humberto told me a bit about his former fortress life in Venezuela.

I had heard a lecture by Aaron Klug[198] (Figure 49) at Columbia University on 3D reconstruction from projections (where I postdoced with Cyrus Levinthal[199] (Figure 50) on *Daphnia* brains). I came up with simpler stochastic mathematics for reconstruction from projections, based on the Japanese art of raking gravel[200,201], probably learned about from a Japanese

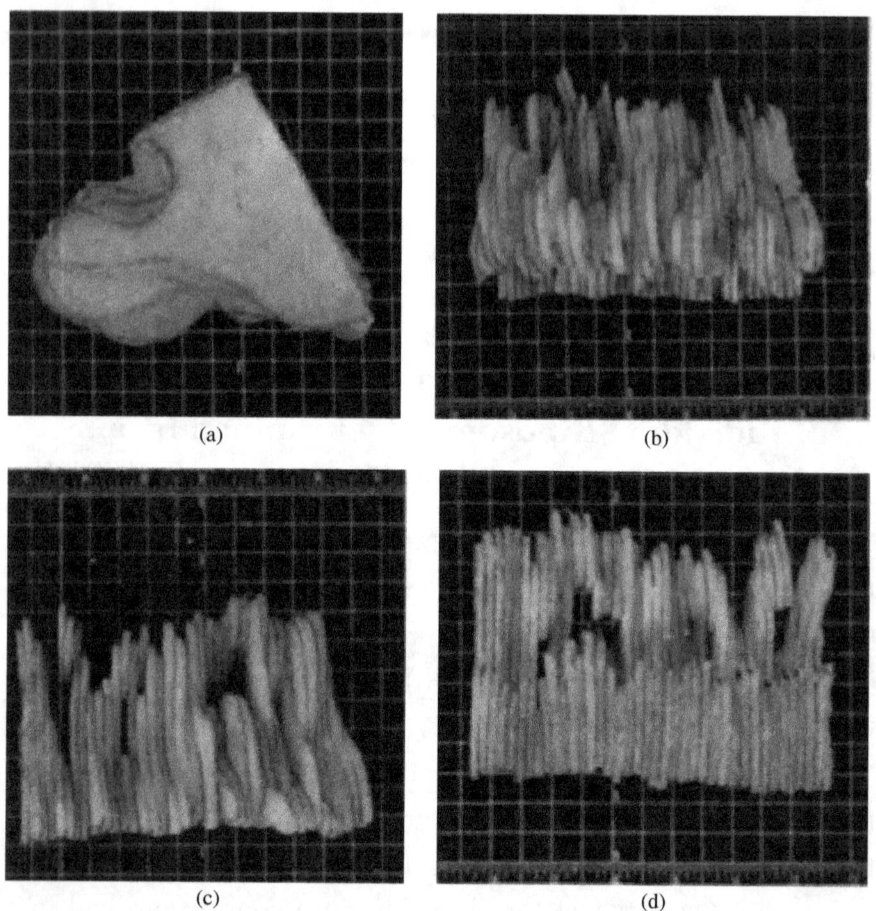

Figure 46.   Styrofoam model of a ribosome[193]. (With permission of Elsevier.)

artist visiting the Art Institute of Chicago, that was also more general[202], which I later realized could be averaged and made simpler. Robert and I joked around with acronyms. As my mother, Diana Gordon (Figure 10), was an artist, I coined the more respectful ART = Algebraic Reconstruction Technique[203], dropping the leading F = fast. It transliterates in Russian, *APT*.

The ±30° tilt range of the electron microscope stage forced me to deal with underdetermined equations, and resulted in the "missing sector in Fourier space".

Figure 47. Humberto Fernández-Morán (1924–1999)[195], at an electron microscope. (Under the Creative Commons Attribution-Share Alike 3.0 Unported license.)

Figure 48. Robert Bender. (With his permission and that of Maclean's[197].)

Figure 49. Aaron Klug (1926–2018)[198] in 1979. (Under the Creative Commons Attribution-Share Alike 3.0 Netherlands license.)

Figure 50. Cyrus Levinthal[204] (1922–1990). (CC-BY 3.0.)

We were attacked in a paper to Jim Danielli's (Figure 35) *Journal of Theoretical Biology*. As editor and still angry at the treatment of Maurice Wilkins (as I recall)[205], he gave us a month to reply, so that Klug's later watered-down attack could be published with our rebuttal in the same issue. As Klug's original article was submitted as a blue ditto (invented before xerox), and was obviously widely distributed as a preprint, our rebuttal quoted from the original[206]. Our lead author was the character, S. H. Bellman, the protagonist in Lewis Carroll's poem *The Hunting of the Snark*[207]. In between[208] and[193] was a blank page that we attributed to Bellman:

> "He had bought a large map representing the sea,
>   Without the least vestige of land:
> And the crew were much pleased when they found it to be
>   A map they could all understand."[207]

The counter attack came from a member of Klug's lab[209], claiming that ART didn't work. His SIRT (Simultaneous Iterative Reconstruction Technique) presently has 1,050 citations in Web of Science, whereas ART[208] has 1,955, or 1866 versus 3876 respectively in Google Scholar (as of Y2024/M10/D16). (Too many for me to read, and mathematically often over my head. Some ART papers may be Adaptive Radiation Therapy, seemingly coined 1997. Selective Internal Radiation Therapy goes back to 1947. Thus, I'll leave it to others to sort this out this medical abbreviation confusion.) Citations of both have taken off since the early 2000s (Figure 51). Many variations of both algorithms have also been published. Between the two papers, the superiority and flexibility of algebraic, iterative approaches was established over Klug's restrictive Fourier Backprojection[210], with its complex interpolation. In[206] we reconstructed Klug's face and the face of a beautiful young lady who had flustered him, who Robert was then dating, and came along for an earlier meeting with Klug. The net result was that we delayed Klug's Nobel Prize by a year, while we dreamed of getting one. (I have since been nominated, in 2022, one amongst 3000 nominees, but for my work in embryology, which others are

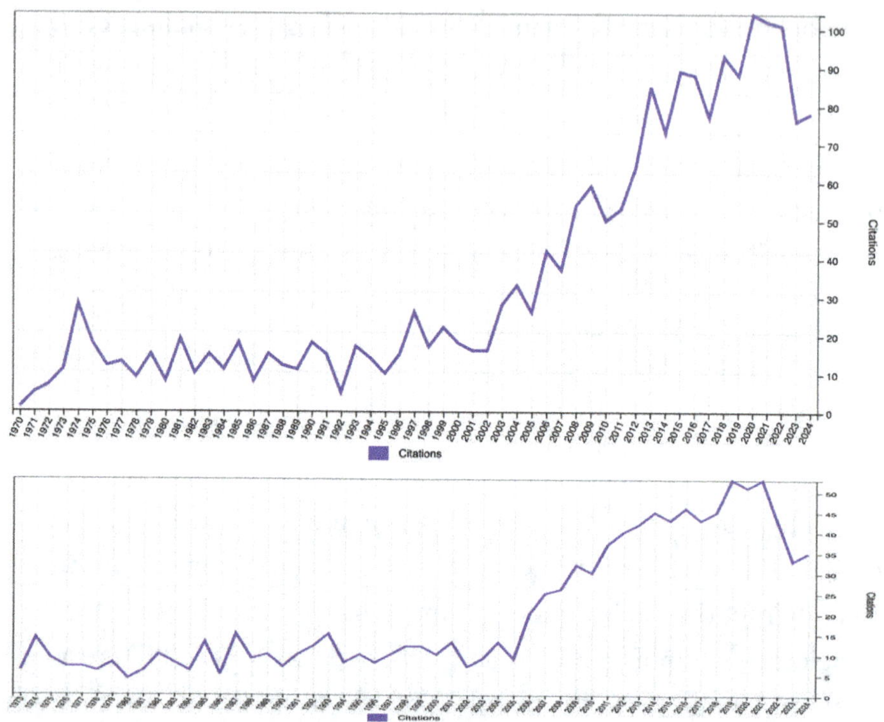

Figure 51.   ART citations (above) and SIRT citations (below) per Web of Science, SIRT rescaled vertically for comparison.

just considering, as for butterfly spot patterns[211]). Our 3D ribosome image was crude[193], as the destruction of the ribosome by the electron beam was obvious in consecutive electron micrographs, and is now just of historic interest as the first attempt. We rented a drum scanner for $1000/month to digitize the film electron micrographs. Some people were taking ribosomes apart, but our letters to the effect that 3D reconstruction of these various partial ribosomes might help solve its structure went unanswered (this was before e-mail and when long distance phone calls were expensive for us two). Another missed Nobel Prize[212].

An amusing episode was that Robert and I cut out a stack of Styrofoam shapes representing sections of the ribosome, decided they were stacked in the wrong order, and in our exhausted sleepiness glued them back together in the same wrong order. Robert was then a chain smoker, and I learned about the strength of addiction when I found myself unconsciously bobbing my head into his smoke.

I briefly got involved in the field of stereology[213,214].

Mathematicians tend to ignore the nonlinear aspects of ART, thereby equating it with Kacmarz's linear algorithm[215]. I discussed the relationship in[201]. MART (Multiplicative ART)[216] is not only nonlinear, but confines the reconstruction of a single object to a convex polygon containing its convex hull[217,218]. The Fourier and Kaczmarz algorithms spread the projection data over the infinite plane. Glen Colquhoun (Figure 52), a retired engineer in Winnipeg and Scottish dancer, contributed to the MART paper[216] and other novel approaches to computed tomography[216,219–222].

At the time, x-ray crystallography was the only way to get the 3D structure of macromolecules[203]. We showed there was another way, which much later led to a Nobel Prize for Joachim Frank[223], who I tried working with on thick sections of salamander embryos

Figure 52. Glen Colquhoun. (With his permission.)

taken with a high voltage electron microscope, in the hope of reconstructing them in 3D: "Joachim Frank made the technology generally applicable. Between 1975 and 1986 he developed an image processing method in which the electron microscope's fuzzy two dimensional images are analysed and merged to reveal a sharp three-dimensional structure"[224]. In retrospect, I've met and provided ideas to a few Nobel Prize winners before they got the Nobel Prize. Is this because I didn't do the experimental work, or because theory in biology never attained the social status of theory in physics? I realized as a young man that those who wander into the wilderness of science are rarely recognized by the settlers who follow, except sometimes posthumously. See the cartoon in[201]. For example, when a scientist from a medical imaging company gave a talk on an expensive machine he was installing, he started out by stating that he was honored to have me present. Not one of my colleagues had any idea what he was talking about.

I read and got context and perspective from Thomas Kuhn's *The Structure of Scientific Revolutions*[225] (Figure 53). We exchanged letters in the 1960s, in which he was surprised that anyone was still interested in his book!

Figure 53.   Thomas Kuhn (1922–1996)[226]. (UC.)

7. Self-sorting of Isotropic Cells (1970)[227]

I met Bob Rosen (Figure 16) when he gave a talk in Colorado on morphogenesis, and he invited me to join him in Buffalo as a postdoc. From him I learned intellectual independence. His abstract attempt to define life occurred after I left him[228]. His earlier books, such as[229], were less abstract and closer to my interests.

This article[227] (Figures 54, 56) was a group effort for about a week at a workshop. It was my first attempt at rheology in 2D[230],

Figure 54.   From[227] with permission of Elsevier.

both later republished in a book[231,232], to which I contributed the computer simulation. (One participant was Richard D. Campbell, who worked on *Hydra*, for whom I caught Canadian cold water *Hydra* to ship to him, while canoeing in Whiteshell Park, Manitoba.) It was an attempt to simulate the onion structure that came from Malcolm Steinberg's[233] (Figure 57) approach to embryonic cells that were enzymatically dissociated and then clumped together[233] in 3D. Over a few days the cells rearranged themselves. Empirically a transitive relationship was seen: if cell type A surrounded B and B surrounded C, then A would surround C. "Onion patterns" were attributed to differential adhesions by a speculated mechanism, leading to an immiscible liquid model for these cell movements[230,231,234,235]. I first got interested in this problem after hearing a lecture by Aron Moscona[236] (Figure 55) while I was an undergraduate student at the University of Chicago[237], which began my interest in embryogenesis. Moscona emphasized the specific surface proteins of the embryonic cells, leading to less elegant modelling[238]. I'm not sure if the two viewpoints ever got

Figure 55. Aron Moscona (1921–2009)[236]. (With permission of the University of Chicago Photographic Archive, [apf1-10586], Hanna Holborn Gray Special Collections Research Center, University of Chicago Library.)

Figure 56. Narendra Goel. (With his permission.)

Figure 57. Malcolm Steinberg (1930–2012)[248]. (From[249] with permission of Princeton University, Denise Applewhite.)

harmonized[239,240]. Reconciling these two models may prove to be important for understanding embryogenesis. Physicists are now doing detailed analyses of clumps of one kind of cell[241]. At least one physicist has considered two cell types[242]. Perhaps we will soon see three cell types yielding onion structures[243]. The ability of *Hydra* and sponges to reform themselves from random clumps of many kinds of cells has always intrigued me[244–247]. We have a long way to go, but this may be a route to whole embryos.

Malcolm had inherited a centrifuge microscope from E. Newton Harvey[250]. The added force changed onion structures in a way compatible with Malcolm's "differential adhesion hypothesis"[234,235,251].

8. On Monte Carlo Algebra (1970)[252]

Terrell Hill (Figure 26) introduced me to networks of reactions, and I in turn introduced him to Monte Carlo computer simulation

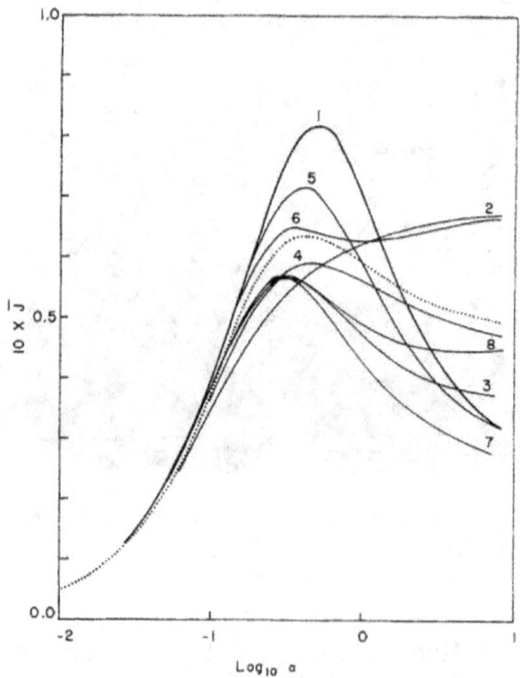

Figure 58.   The dashed line is the precise calculation for a membrane channel with 8 sites. Most of the Monte Carlo samplings of terms capture the form of this curve[252]. Flux versus rate from a bath to one side α. (With permission of the Applied Probability Trust.)

of those networks[121]. One problem that intrigued me, from my studying traffic flow, was the problem of crossing a one-lane bridge, with traffic going both ways. My thesis work on steady state flow across a membrane within which the moving particles attractively interacted, causing a 2D Ising lattice phase transition that blocked transport across the membrane[253] led me later, at Columbia University, while postdocing with Cyrus Levinthal[199] (Figure 50), to narrow that model membrane to a single row across it, to what might be called a "membrane channel". I considered the linear equations equivalent to this steady state flow across the channel, similar to the one-lane bridge. They got more and more complicated with the thickness of the membrane. The equations had

Figure 59. Airplane contrails over Alonsa, Manitoba, making polygons in the sky. Light blobs are reflections of lights from inside the window through which this was taken for me by Frank Chen, not UFOs. (Reproduced with his permission.) This is analogous to the polygons used in RBYRCT (Ray By Ray Computed Tomography)[136].

more and more terms. So, I came up with the idea of assuming that the terms might have similar values, and took a Monte Carlo sampling of the terms. Now those terms selected at random were still in algebraic form. Thus, I called this symbolic computing "Monte Carlo algebra", implying that the solution of linear equations could have this stochastic component. I used the thickest membrane for which I could still do the exact calculation, to provide a comparison with the Monte Carlo samplings when numerically evaluated. I learned Formac, a computer language that preceded Mathematica with its much more sophisticated symbolic computation capabilities. Formac generated huge Fortran programs on punched cards. The runs required special settings on the mainframe computer at Columbia University, so I ran them overnight with special permission. I recall my chagrin when Columbia students tried to attack the computer center, and were stopped by black university guards.

As Monte Carlo algebra was, and still is, a most unusual use of probability theory, I so to speak decided to go into the lion's mouth, and submitted it to the editor of the *Journal of Applied Probability* and author of[254]. The paper was accepted. While computers are much faster now with more memory, the Monte Carlo algebra approach waits to take on more problems, perhaps a step towards quantum computers.

I am now working with others to find algorithms for ray-by-ray computed tomography (RBRCT)[136], an entirely new approach to computed tomography. These indeed proceed from a sampling of rays directed across an object (the human breast in our case), and have the potential of much dose reduction compared to fan beam and cone beam approaches. The only analogs are cross-bore computed tomography, used in geology, and trying to image whatever is inside a wall from both sides. RBRCT is similar to aiming a rifle, whereas fan and cone beams are similar to shot guns, the latter indiscriminately irradiating normal tissue. But we're trying to be more clever, aiming a narrow X-ray beam, delivered by a parallel optics polycapillary X-ray source[255], to minimize the X-ray dose, not envisioned in our massive review[256] (Figure 60). The aiming is dependent on the estimated image at each stage. While the polycapillary x-ray source is proposed to be steered by an industrial robot in the first simulated prototype[136], faster electronic steering may be possible[257].

Xiaohua Zhou, then a graduate student (Figure 60), had accompanied me as translator on a UN sponsored trip to China, where I lectured for a week in Chengdu. While there, the young lady he wanted to marry turned him down. A year later, she changed her mind and came to Canada. Natalie and I loaned our house to them, and occupied her parent's cottage for a while.

I was short of grant money for a while, so Xiaohua took a job with the head of Medical Physics, who at that time was culturally a UK Israeli, who shouted at him, contrary to Chinese tradition. Xiaohua was at a loss what to do. Natalie gave him lessons at shouting back. Xiaohua and Shlomo Shalev worked together well from that point onwards.

Figure 60.   Xiaohua (Albert) Zhou and me. (With his permission.)

9. A Capillarity Mechanism for Diatom Gliding Locomotion (1970)[258]

I was hooked on morphogenesis from my earlier experience with simulating a 2D snail[79]. But I got sidetracked. On reading[260], reprinted from the 1952 original by John Tyler Bonner, I was introduced at the beginning of this book on morphogenesis to the remarkable colonial diatom *Bacillaria paradoxa*, in which the individual diatoms actually slide back and forth against each other, a motion not seen elsewhere in the living world. Thus began my substantial career in trying to understand diatoms, mostly their morphogenesis and motility[261,262]. *Bacillaria* was amongst the first diatoms discovered, back in 1783[263]. It has puzzled everyone since then[264]. It's best understood if the reader watches at least one of the many *Bacillaria* videos online.

I was invited by a former postdoc, Rangaraj Rangayan (Figure 125), to speak at an engineering conference[265]. Not knowing

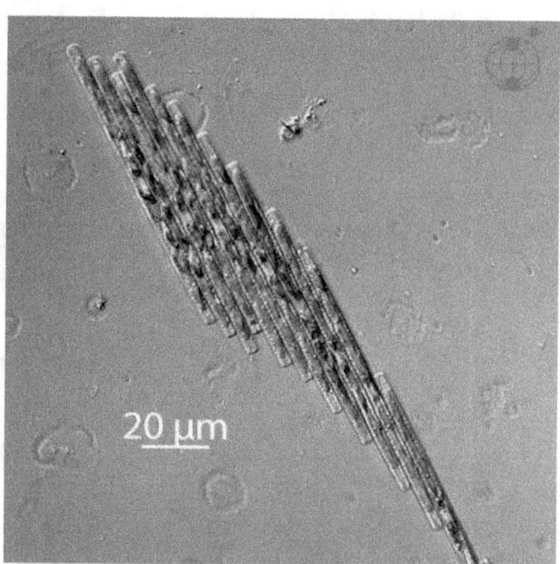

Figure 61.   *Bacillaria*[259]. (Permission: Creative Commons Attribution 3.0.)

what would be of interest to engineers, I wrote the first paper on diatom nanotechnology, a field which has flourished since. Probably, most authors have independently discovered the usefulness of diatoms for nanotechnology, so I won't claim that I kickstarted the field. Afterwards I only made small contributions to diatom nanotechnology[266–268].

The motility problem for *Bacillaria* and other, single-celled diatoms, addressed in[258] with an analogy between a diatom and a candle flame when I lectured on it, continued to plague me, until I came up with a complex model involving concerted motion of raphe fibrils and their membrane synthesis[269,270] (the latter[270] with the superb programming of Shruti Raj Vansh Singh (Figure 62) and Krishna Kaytal (Figure 63), after reviewing the many, sometimes outlandish models over the previous centuries[271,272]. (Mine may end up in that heap.) Earlier, with the help of electron microscopist Don Parsons[273] and Robert Bender (Figure 48), we had

Figure 62. Shruti Raj Vansh Singh. (With her permission.)

Figure 63.   Krishna Kaytal. (With his permission.)

destroyed *Bacillaria* cell #2 with a pulse ruby laser aimed through a light microscope, and noted that cell #1 went back and forth in an FM wave (rather than AM) against a shard of cell #2[45,274]. This proved that the oscillatory motion was intrinsic to each cell. But their coordinated motion remained unexplained[275]. I had an excellent botany graduate student, Margaret Gowdar (nee Kapinga), who observed their behavior[276–280]. For instance, she found they were not diurnal, but moved any time the light was on. Since their resting stage in the dark involved all cells aligned like a deck of cards, I postulated that light piping might occur along a colony to synchronize their movements[268,281]. A full understanding is still being sought[262,264,282,283]. Light piping in any colonial diatom chain remains to be tested, perhaps by using an optical fiber to put light in at one end and observing where it comes out. Light piping might be used to distribute available light to all cells in a colony[268,284]. Stan Cohn (Figure 64) discovered that pennate diatoms have spots

Figure 64. Stanley Cohn. (With his permission.)

Figure 65. Wayne Brodland[301]. (UC.)

at the end of each valve that are light sensitive, with a spectrum different from photosynthesis pigments[285], and reverse direction when entering a dark zone. I guessed that in *Bacillaria*, alignment of these spots effectively creating a transitory light pipe may be related to the colony's resting configuration and/or synchronization. I visited a lab that makes complex, periodic glass light pipes[286].

Wayne Brodland (Figure 65) and I analyzed a sharp square in a diatom shell and concluded it was due to a salt crystal inside a diatom silicalemma during its valve morphogenesis[287]. John Parkinson (Figure 66), then a postdoc with me, extended the DLA (Diffusion Limited Aggregation) model that roughly explained the morphogenesis of diatoms in[288] to include a plausible role for microtubules[289]. Frank Round (1927–2010)[290], editor of *Diatom Research*, sent me a personal note praising my viewpoint of diatoms. I continued modelling diatom morphogenesis with Vadim Annenkov[291–293], and became a series editor for books on diatoms with Joseph Seckbach for Wiley-Scrivener[87,150,293–298], but perhaps due to the unique ability of diatoms to produce species specific 3D structures over 8 magnitudes of size out of a single substance, amorphous silica[268], there is much yet to be understood[87,295,299,300]. Diatoms, being single cells with incredibly ornate silica shells, might lead eventually to an understanding of the relationship of the genotype to the phenotype, lack of which currently plagues all of biology, including evolution. Certainly, the morphogenesis of single cells, such as diatoms, radiolarians and ciliates, deserves

Figure 66.   John Parkinson. (With his permission.)

much attention. It is generally ignored by those who study the morphogenesis of multicellular organisms.

10. Development of the Heterocyst Spacing Problem in *Anabaena* (1970)[302]

What used to be called "blue green algae" in a turf war between botanists and microbiologists, are now referred to as "cyanobacteria". The microbiologists won. I ran across them in keeping my eyes open for simple morphogenesis problems, because some genera such as *Anabaena* are 1D chains of vegetative cells. As the vegetative cells divide, generally nondividing cells called heterocysts form singly or in pairs (they fix nitrogen, turning inert atmospheric nitrogen = N_2 into compounds used by all organisms, essentially nitrogen fertilizer). As two heterocysts become further apart due to cell divisions of the vegetative cells between them, one or two vegetative cells halfway between turn into heterocysts.

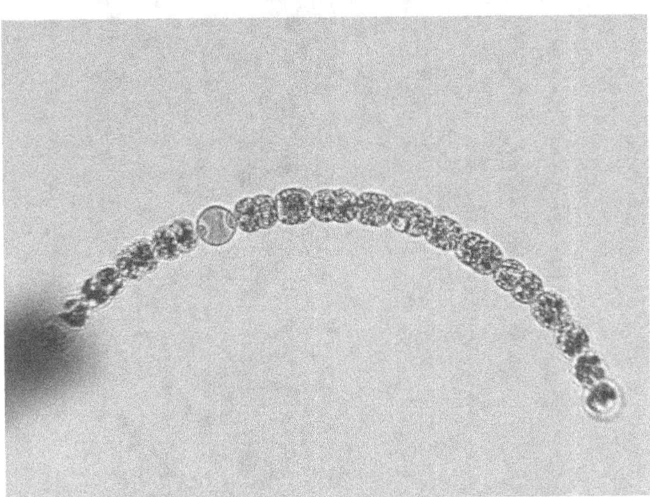

Figure 67. Heterocysts are generally round and spaced with vegetative cells in between. This micrograph shows three heterocysts, which are spherical. Two are at the ends[303]. (CC BY-SA 3.0.)

Thus, a simple "spacing pattern"[304] is formed, but one that is just 1D. Enamored of this, I collected about 1,000 papers on cyanobacteria, learning many nuances. I had tried a 1D simulation of *Hydra* while at Columbia University, which was never satisfactory, but had done a better simulation of heterocyst spacing patterns, as acknowledged in[305]. When I met Gabor Herman (Figure 68) later at SUNY/Buffalo, where I was a postdoc at the Center for Theoretical Biology and he was an assistant professor in Computer Science, I found him working on problems I had discussed with him. In my lower (postdoc) position, I decided to collaborate. However, in regard to heterocyst spacing, the only thing I have to show is acknowledgements by Herman[305,306] and the above conference talk with Tony Walsby. Tony became prominent later after he discovered "square bacteria"[307] (Figure 69), in an Egyptian saltern,

Figure 68.   Gabor Herman, by Andreas Alpers. (With Gabor's permission.)

Figure 69. Tony Walsby (1941–2024)[310]. Photograph by R. A. Lewin. (With permission of the Royal Society.)

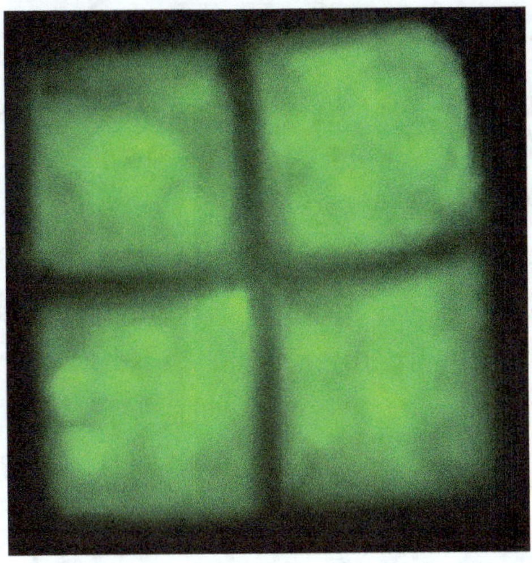

Figure 70. *Haloquadratum walsbyi* micrograph[311], 2 to 5 microns (µm) squares 0.1 to 0.2 µm thick. (Public domain.)

which turned out to be Archaea[308] (Figure 70). Thus, I was quite aware early of the existence of flat, polygonal prokaryotes. Tony also discovered an unexplained phenomenon, that motile colonial cyanobacteria in a flask clump and compact together, while retaining the shape of the flask[309]. This may be somewhat analogous to gravel made from slag (probably quenched blast furnace slag), of which I obtained a sample, a leftover from steel production, each particle with concave surfaces, which intertwine and compact. Concave slags and cyanobacteria clumping deserve more study. Polygonal Archaea later led me to a model for the origin of life[73].

11. New Three-dimensional Algebraic Reconstruction Techniques (ART) (1971)[312]

This extended abstract contains much practical advice on how to reconstruct objects in electron microscopy.

12. Trees (1971)[313,314]

I had done a 2D simulation on a square grid of branching and anastomosing patterns, analogous to the veins in many leaves, while at the University of Colorado, but never published it, probably because I didn't want to think through the general angle computing problem, not amenable to a grid. Later, in Buffalo, Jim Danielli (Figure 35) asked me to review the above two papers, that had

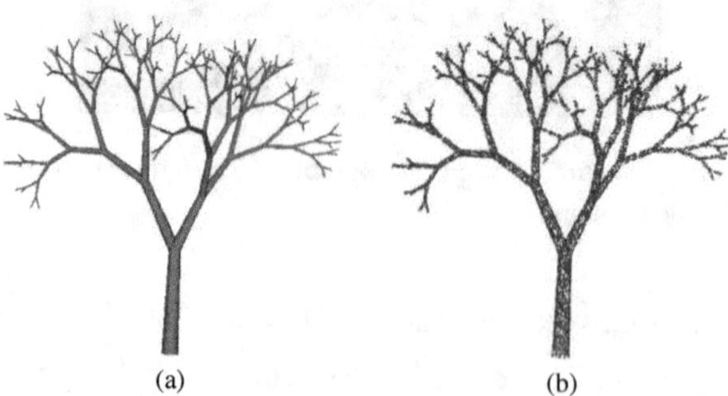

Figure 71. Computer simulations of a tree[315]. (With permission of MDPI.) See also[313,314].

come in about simultaneously. Being a young wise guy, I responded positively with Joyce Kilmer's poem which starts "I think that I shall never see / A poem lovely as a tree"[316]. Branching structures were everywhere, such as river deltas, and ended up in my differentiation trees[183].

13. Reconstruction of Pictures from their Projections (1971)[202]

We were invited to use an experimental computer printer at Xerox, to get our picture output. To accomplish this, Robert Bender

Figure 72. This photograph of a little girl, Judy, lower right, was taken by Judy Carmichael, wife of Jack Carmichael[317]. It was digitized pixel by pixel by hand using a spot digitizer with analog voltage output with a photographic print being moved to 49 × 49 positions. The images were printed at Xerox Computer Science Laboratory in Rochester, New York. [With permission of the ACM (Association for Computing Machinery).]

(Figure 48) arranged to get our magnetic tapes converted to paper tapes at the National Research Council Canada in Ottawa, where I recall his father worked. We drove there from Buffalo in his beat-up car with a leaky radiator, which seemed to require 5 gallons of water per mile. We then took the paper tapes down to Rochester, New York, where we were allowed to use the Xerox image printer overnight, unsupervised. I think "Judy" was the first digitized image of a black person. In my previous attempts to find a digitized image while I was at Columbia University, I could only find NOAA (National Oceanic and Atmospheric Administration) digitized pictures of clouds, which were highly unsatisfactory for testing structures. Now, of course, every image on the Internet is a digitized image.

Around this time, Ryan Drum (Figure 86) brought to my attention a toy "car" that required no gasoline (petrol) or electricity. It was powered by compressed air[318], I guessed scaled up to be sat in, enough to get from one urban gas station to the next. At that time, pressurizing your tires was still free. I remembered that when I was a kid, gasoline cost only 9 cents per gallon in Arizona. This was an early lesson in inflation and alternative fuels.

14. Algebraic Reconstruction Techniques (ART) for Three-dimensional Electron Microscopy and X-ray Photography (1970)[208]

I came to the realization that we could substitute the average of projections for the averaging of the images from our stochastic approach[202]. Thus, was the ART algorithm conceived[208]. Publication dates are delayed, depending usually on how fast reviewers respond, and so the chronology gets slightly mixed up.

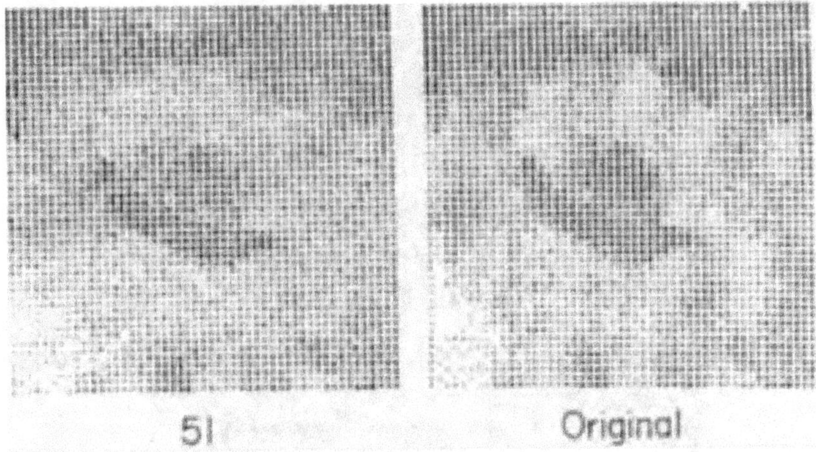

Figure 73. Reconstruction of the image of a lunar crater from 51 views over ±30° from the horizontal demonstrated that the nonlinearity of the ART algorithm made up for the missing sector in Fourier space. (With permission of Elsevier.)

15. Saltation of Plastic Balls in a 'One-dimensional' Flume (1972)[191]

Jack Carmichael[317] was a postdoc with Terrell Hill at the University of Oregon, while I was Terrell's graduate student. I got fascinated by the streaks of color as the waves washed over Oregon's beaches and their tidal pools, which I camped nearby with an adventurous blind student Dan, a would-be scientist who, tinkering with silver nitrate as a teenager, exploded off both of his hands and lost his vision. He aspired then to become a social worker. Jack was working on the theory of chromatography, a method for separating different kinds of molecules from a mixture, so I called the beach phenomenon "the self-chromatography of sand". Jack invited me to join him for a summer, later, at the University of Massachusetts, in the Department of Polymers, where I got to hear a series of lectures on polymers by Paul Flory[319]. Jack got intrigued, and we

Figure 74.   Strobe pictures of a ball saltating over a 1D bed of balls. Water flow is left to right. Ruler is in centimeters. (From[191] with permission of John Wiley and Sons.)

constructed a long, narrow plexiglass flume, long enough to watch small plastic balls denser than water bounce over a bed of the same balls, and photograph them with a strobe light[191]. Later, at Columbia University, I wrote a computer simulation of the process. We showed that the roughness of the bed was sufficient to maintain the long-term repeated saltation of the balls, while saltation damped out rapidly on a smooth surface. Probably after publication, I visited Sir Ralph Bagnold[320,321] (Figure 75), who has had a Mars dune named after him[322], at his retirement home in rural England, who during World War II on military duty in northern

Figure 75. Ralph Bagnold (1896–1990). (Under the Attribution-Share Alike 3.0 Unported — CC BY-SA 3.0.)

Africa spent much of his time studying saltation in sand dunes. Bagnold was inspiring and supportive. The one-dimensionality of the flume harkened back to the one-lane bridge, and afterwards to 1D models of the diatom raphe involved in its motility. So, even though we published it in *Water Resources Research*[191], intellectually it was not a great deviation. In retrospect, this ability to work in many disciplines and see relationships between them has been both important and fun.

16. Towards a Theoretical Biology 4. Essays (1972)[323]

During my stay in Jim Danielli's (Figure 35) Center for Theoretical Biology, SUNY/Buffalo, I postdoced briefly with visiting Conrad Waddington[327] (Figures 76, 78), known for his forays into embryology[328] and art[329]. While I was primarily interested in embryology, attending an embryology course by Reed Flickinger[330], Waddington had moved on to futurism, and had brought two young fellows with him, so inclined. Nevertheless, he invited me to his fourth conference on Theoretical Biology at a Rockefeller mansion in Bellagio, Italy. Unfortunately, I didn't pay attention to the publication deadline, so all I have to show for this privilege is the group photo (Figure 76), and a few memories: of the cab driver who scraped his car against some mountain that had a road *en route*, cut into it, discussing the number of petals on a daisy privately with Francis Crick, excellent piano playing by Christopher Longuett-Higgins, and my first experience with formal dinners. Lewis Wolpert, who I visited for a month in London, and Stuart Kauffman were there, whose paths I continued to cross. Both inspired me to consider 1D

Figure 76.   Lewis Wolpert (1929–2021) is at the far left, top, with me adjacent[323]. Stuart Kaufman[324] is third from right, top. Conrad Waddington (1905–1975) is fourth from left, bottom, Christopher Longuett-Higgins (1923–2004) third from left, and Francis Crick (1916–2004) is third from right, bottom. (Former Edinburgh University Press, UC.)

models, Wolpert: *Hydra* morphogenesis, and Kaufmann: lines of animals, such as men spacing themselves for privacy along a row of urinals, or birds clustering along overhead wires. Waddington's (Figure 78) "epigenetic landscape"[326] (Figure 77), which I confess I never understood, later became instantiated as my "differentiation trees"[183,331,332]. What is a ball, what is gravity in this metaphor, how is a decision made at a bifurcation, what about multiple cell types (balls?), do balls split? These mysteries kept me

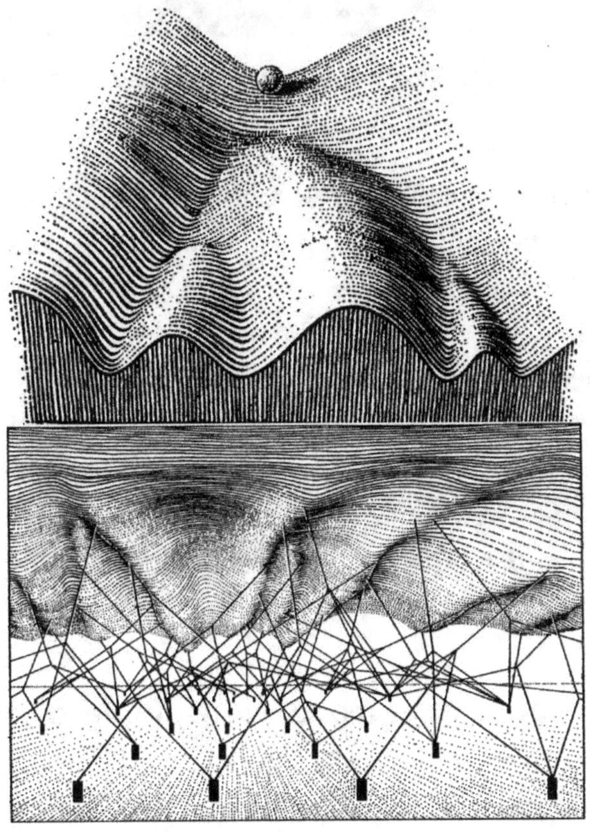

Figure 77. Waddington's epigenetic landscape, seen top and bottom[325]. The balls somehow represent cells differentiating in an embryo. The shape of the landscape is controlled by the genes below. I never understood it. (Cf.[326].) Figures from[325]. (With permission of Taylor and Francis (Books) Limited UK and Informa UK Limited.)

Figure 78.   Conrad Waddington (1905–1975). (With permission as "a free resource"[343].)

from swallowing the metaphor. It took decades of thinking around Waddington's metaphor to observe and realize the possible importance of differentiation waves[149,183,333], the possible existence of a differentiation code[334], and the Janus-faced nature of causality in embryogenesis[335,336], perhaps requiring a new look at reductionism[337]. Waves are an example of global/local interactions. There are also ramifications for macroevolution[183,338]. The relationship of differentiation waves to the genome[339] is yet barely understood. As waves occur in many parts of organisms[340,341], including plants[342], we might expect a breakthrough in understanding them.

17. Artifacts in Reconstructions made from a few Projections (1973)[344]

At the time, there was concern about adding complicated constraints to the linear equations for reconstruction from projections (now called "computed tomography"), presently effectively addressed via Artificial Intelligence (AI)[345–347]. In terms of the abstract concept of the hyperplane of solutions, all iterative algorithms seemed to me to converge to pretty much the same solution, in that the converged images looked similar. I came up with the idea of intelligently exploring the hyperplane of solutions to get away from this common region[327] of hyperspace. The ART algorithm has the property that from any point in hyperspace, it would converge to a point on the hyperplane of solutions. Thus, by adding or subtracting objects to a reconstructed image, one could project that image onto the hyperplane of solutions and see if the object survived the process. It either did survive or didn't survive, either way providing what I called an "intelligent walk in the hyperplane of solutions" (Figure 79).

This approach, albeit tedious, worked, and I had a colleague give me projections of a picture not known to me. I was able to partially recover its features. I dimly recall being asked to submit this conference paper to a journal[348], but being a young purist, I would not then publish the same article twice in the scientific literature. Ironically, IEEE lost all of their records of this conference, so my personal copy is the only record. I plan to copy it in[136].

If one could find a metric that made a grid over the whole hyperplane, each point on that grid would represent an alternative solution to the equations. Thus, if the corresponding reconstructed images were placed before one, the true object would be amongst them. How one recognized the true object depended on highly nonlinear information. For instance, in electron microscopy, if one knew from the chemistry that one was dealing with a single molecular object, then any reconstructed images that showed two or more separated parts could be rejected. The problem was that for decades I puzzled what that metric could be?

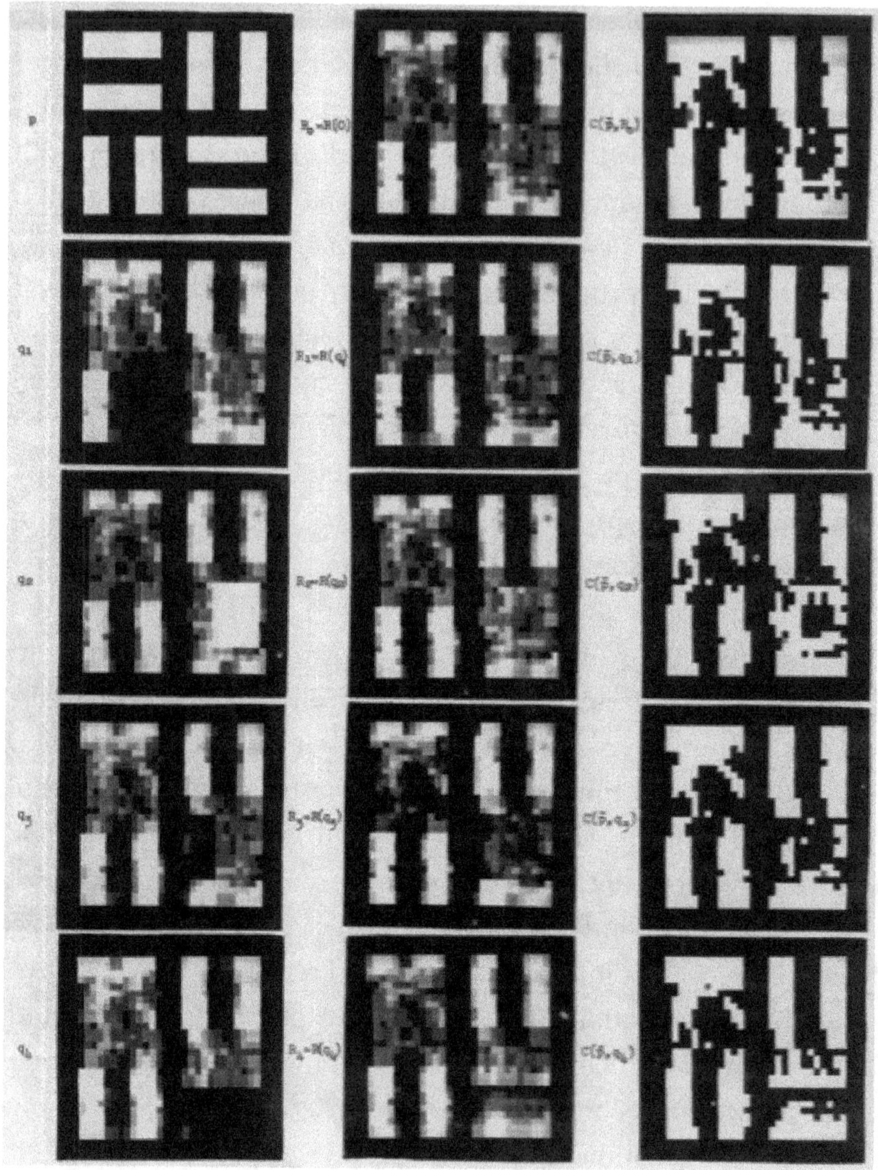

Figure 79.   Was it there or not? Adding and subtracting rectangles to a reconstructed image, then converging that image to the hyperplane of solutions[344]. The original is in the upper left corner, and was not seen until the other images were done. (With permission of IEEE.)

Much later I came up with the idea that any basis function might provide such a grid. Basis functions are countable, and finite in number for any picture represented as a finite number of pixels. Each basis function projects to a point on the hyperplane, effectively creating a grid across the hyperplane. A paper was written with Chengxiang Wang in China[349] (Figure 80). Optimal choice of a basis function still needs to be explored. We tried a few.

One might think that a solution such as ART would be of sufficient value, but the prospects for improvement over ART have always been tantalizing, albeit sometimes requiring such abstract forays. The richness of the problem of solving underdetermined equations has always fascinated me.

An example is our new approach to the MART algorithm in[136]. The rays are considered individually, instead of in bundles, such as in 2D fan beam CT or 3D cone beam CT. Consider a given ray. Let's make a graph in which each ray corresponds to a node. It crosses rays that could be considered adjacent in the graph theoretical sense. Edges in the graph could reflect this. In turn, those

Figure 80. Chengxiang Wang. (With his permission.)

rays cross others. Now suppose that we apply the MART algorithm only to those rays within graph adjacency $i \leq n$, where n is the current number of lines. That might be adequate for small i, if the local nature of CT algorithms[350] applies to this graph[344]. We are exploring this possibility, which would vastly speed the computation if it works. In a sense, this graph is the dual of the pick-up sticks graph of the rays. The spaces between rays can be filled in with an inpainting algorithm, as shown with fresh high school graduate Ken Chen (Figure 82).

In another approach, collaborators Alexander Konolov and Vitally Vlasov (Figure 81) have developed an algorithm that initially generates the minimum number of X-rays so that for a given diameter, all tumors are initially intercepted by at least one X-ray.

Figure 81.   Alexander Konolov (*right*) & Vitally Vlasov (*left*), in 2019. (With their permission.)

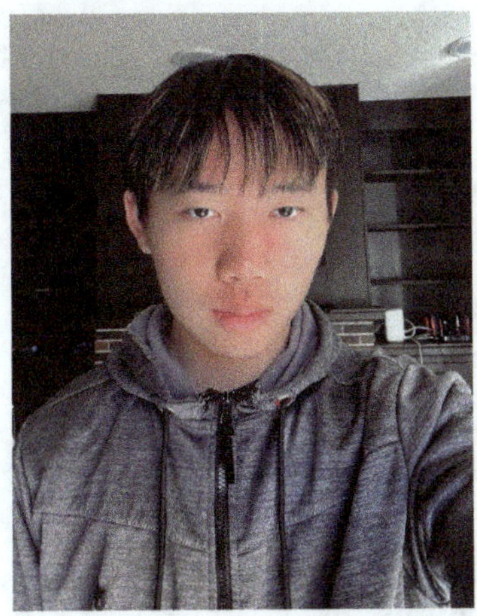

Figure 82.   Ken Chen. (With his permission.)

18. An Alternative View of Morphogenesis (1974)[351]

An attempt to smooth the fine branches by simulating sintering proved too slow for computers at that time to explore branching (Figure 84). Faster computers have allowed an alternative model to DLA with sintering to be considered[352], though the DLA with sintering model[288] has yet to be tested for branching and compared to it.

This was my first argument against the then prevailing model of morphogenesis, Alan Turing's reaction/diffusion model[353]. In subsequent work for Turing's centenary, I showed that even if the Turing model applied to one step of differentiation, it could not handle a second step[354]. Turing's ideas have been dominant since his last paper with his reaction/diffusion model for morphogenesis[353], which did not seem correct to me, in terms of marine angel fish pigment patterns[183], ability to explain more than one step of differentiation[354], and gradient models[355]. Biologists receive little mathematical training, and are thus often innumerate[356]. For

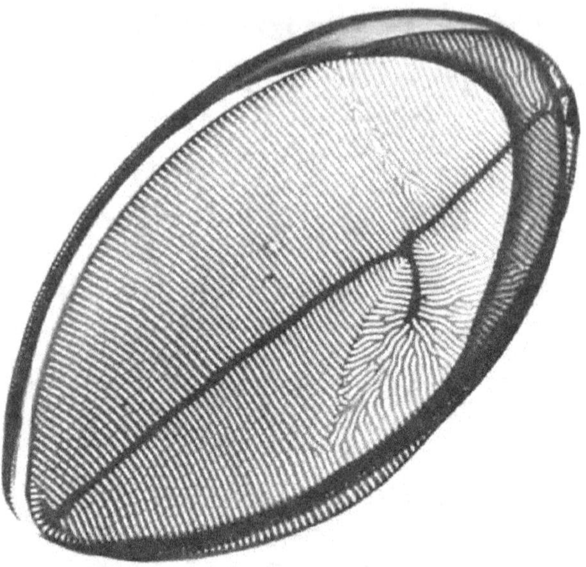

Figure 83.   Electron micrograph of an aberrant *Licmorpha* diatom[288]. (With permission of Elsevier.)

Figure 84.   Centric diatom DLA (Diffusion Limited Aggregation) simulation[288]. (With permission of Elsevier.)

instance, the speaker at one seminar claimed that two perpendicular gradients of the same hypothesized substance could provide two coordinates for a cell in the positional information model. I pointed out to his confusion that only a single diagonal gradient would occur, yielding but one coordinate.

The paper for this section was an obscure publication[351] with the first announcement that we planned to tackle the morphogenesis of diatom valves as analogous to branching frost patterns. I read a lot about dendrites in crystallography. The lead editor, Aharon Katchalsky, was later killed by a terrorist from Japan while in the Israel Ben-Gurion airport. Had he lived, I might have postdoced with him and become an Israeli. The work came to fruition two decades later, with the introduction of a DLA (Diffusion Limited Aggregation) model for diatom valve morphogenesis[288].

Botanist Ryan Drum (Figure 86) from the University of Massachusetts, whom I met while there in 1968 and again while attending the Woods Hole Marine Lab embryology course in 1969, then an excellent electron micrographer of diatoms[357], provided a great source of key image data, especially for diatom motility[258]. We wrote a review of diatom morphogenesis that introduced diffusion limited aggregation (DLA) as a first approximation[288] and discussed diatom nanotechnology[358]. With his eccentric ways, he did not last in academia, and became a herbalist[359]. For example, when I invited him to talk to a group of undergraduates at SUNY/Buffalo, he spoke sitting on a toilet seat while his movie of a drive across the USA compressed to 1 hour played in the background: ground level at 3000 miles/hr. I have 100 electron micrographs of aberrant diatoms of the genus *Licmorpha*, that he took in the 1960s, waiting for analysis and possibly computer simulation. When I visited him at his subsistence farm on Waldron Island, Washington, no human presence was obvious near the empty field on which I was deposited, except for a sign "Beware of low flying planes". He showed up ½ hour later.

We wrote *The chemical basis of diatom morphogenesis*[288,360], taking the title in part from the classic *The chemical basis of morphogenesis*[353], Alan Turing's (Figure 85) last publication before the United Kingdom perhaps inadvertently killed him with drugs to "cure" him of being gay, even though he had cracked the WWII German's code and designed "Turing machines", what we now call computers, which defined whether or not a problem was "computable". I contributed to a centennial book for Alan Turing[361].

Charles Reimer (Figure 87) allowed me to stay overnight at the Academy of Natural Sciences of Philadelphia, to use his card catalog on diatoms, in which I discovered key papers going back to the early 1800s that using vapor deposition of silica as an analog supported the DLA view of diatom valve morphogenesis. Ryan and I also made a plausible model for diatom motility[258], published again in PNAS with Terrell Hill's backing, regarded as an obscure journal for diatomists by Ryan.

Figure 85.   Alan Turing (1912–1954)[362]. (Public domain.)

Figure 86.   Ryan Drum in 2015. From[359]. (With permission of his daughter.)

Figure 87.   Charlie Reimer (1923–2008)[363], (UC.)

The "best" students were asked to stay 6 weeks after the Woods Hole embryology course, but I had centrifuged sea urchin embryos and separated them by degree of mechanical wobble as they ascended a meter long capillary tube, a form of self-chromatography of the pluteus larvae. Not amused by this sacrilege, the lab instructors did not invite me to stay on. A hydrodynamic understanding of wobbling larvae has yet to be done. Eden Miller (Figure 84) worked with me at GSML on the problem, but we ran out of her time there.

Vladimir Annenkov (Figure 214) at Lake Baikal, Siberia, took up the immense task of finding a molecular basis for diatom morphogenesis[291,292] (cf.[364]). As diatoms span 8 orders of magnitude of structure made from a single substance (amorphous silica)[268], the full task may take generations. One approach that may help is cloaking of diatoms, which involves immersing them in a solution whose refractive index matches that of diatom silica[268]. Cloaking could allow the use of modern light microscopy techniques to clearly watch what is happening inside the living diatom cell. For now, the best we have is the superb, mostly bright field movies of the late Jeremy Pickett-Heaps[365] (Figure 240), or attempts to use electron cryomicroscopy[366]. In a winter Woods

Figure 88.   Eden Miller. (With her permission.)

Hole course on *Optical Microscopy and Photomicrography in the Biomedical Sciences* I took with Robert Allen[367] (Figure 89) and Shinya Inoué in 1978, I surreptitiously broke a diatom slide with a salesman, and showed that the diatom valves disappeared from view when covered by immersion oil, which matched the refractive index of glass. I also became fascinated by Nina Allen's (Figure 90) observation that for an as yet unexplained reason, cell nuclei can have quite distinct shapes[368], perhaps a prelude to our Wurfel toy model (Figure 175). Later, in the lab of Kalina Manoylov (Figure 39), I did the same experiment with live diatoms with the same result, except that they died due to the toxicity of the immersion oil. Thus, cloaking works, but requires a nontoxic material of the correct refractive index, perhaps methyl cellulose or its derivatives.

Lewis Wolpert (Figure 76) introduced the concept of "positional information" in 1969[370], using the French flag[355] as an analogy. The idea was that embryogenesis involved gradients of morphogens, and a cell could read the local value of a gradient

Figure 89. Robert Allen (1927–1986)[369], (UC.)

Figure 90.   Nina Allen. (UC.)

and "look up" how it was supposed to differentiate. I played a bit at simulating this model, and eventually concluded that a cell had no idea where it was, but rather that whole groups of cells responded to waves of apical contraction or expansion that passed through them by then differentiating[183,371,372]. The difference between Wolpert's approach and mine was spelled out in[355]. Positional information seemed to me to be similar to ideas of Hans Driesch[373].

I had spent a month at Wolpert's lab in England, mostly with his postdoc, Dennis Summerbell[374], while US student protests against the Vietnam War crescendoed. A highlight was that I found a Pakistani restaurant that saved me from English cuisine. The waiters brought me three pitchers of water at each meal: I thus learned to eat delicious hot (spicy) food. One weekend, lonely, I walked to Wolpert's home, his first visitor ever, a comment on English vs American life. I learned cultural nuances. Wolpert was interested in the small animal, *Hydra*, at the time. Someone sent him a poor copy of a review of a book on Hydra, so he bought it. It turned out to be about the island of Hydra in Greece: a practical joke.

I never directly learned Wolpert's attitude about being contradicted[375]. However, Lev Beloussov (Figure 140) and I prepared a whole session for a Swiss embryonics conference[376], including many speakers, on invitation, and every single paper submitted for our session was rejected by the committee on which Wolpert sat, including our own. I was merely amused, especially when the organizers tried to insist that we attend anyway. (Embryonics is the attempt to build computers that act like embryos. I liked the idea, but thought they used the wrong model for embryos[377–379].) At any rate, we didn't go. This rejection tradition of outsiders goes on, as my recent voluminous book on the origin of life[73] was rejected when I volunteered for a webinar. One needs a thick skin as an outsider. One also has to be flexible: at a talk I gave to a group of mathematicians only three showed up. Rather than being upset, I sat down with all three around a table, giving them a more personal "talk".

When Lev, a champion for morphomechanics[380], died, I helped out with his memorial[381]. I've honored other deceased scientists this way[382,383].

19. A Tutorial on ART (1974)[384]

I kicked myself for sending this tutorial to a journal that I afterwards learned that the NIH (US National Institutes of Health) Library didn't receive (I worked at NIH at the time), so that I thought it was in an obscure journal. Nevertheless, it was broadly received. Nothing mathematically fancy here, but I did try to be precise.

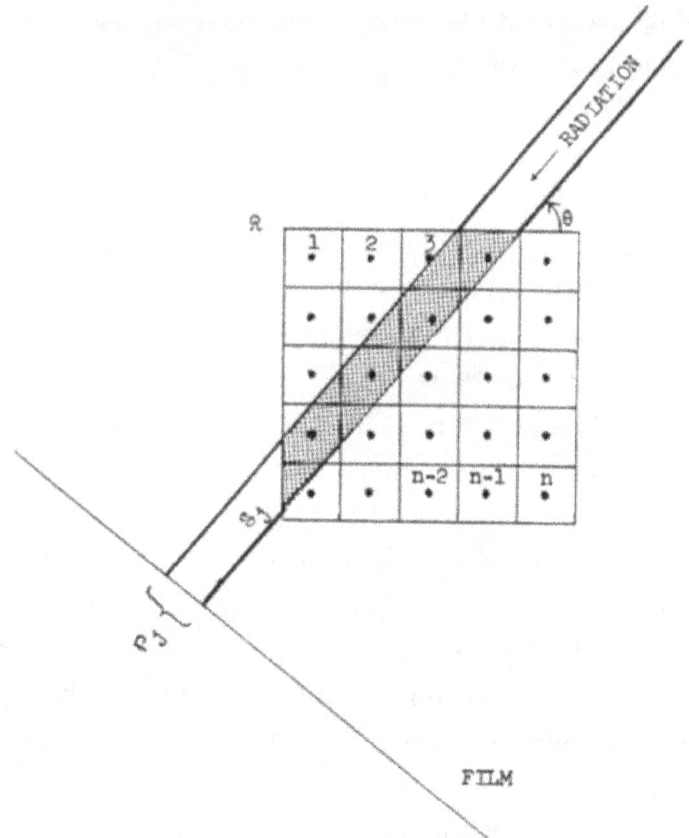

Figure 91.   The general idea of a single parallel ray projected to a line of x-ray detectors)[384]. (With permission of IEEE.)

20. Introduction to the Session on Experimental Aspects of Reconstruction from Projections (1974)[385]

Paul Lauterbur[388] (Figure 92) was the discoverer of MRI (Magnetic Resonance Imaging) initially implimented with my ART algorithm[389]. The two of us organized this session of a meeting held at Brookhaven National Laboratory, the next year[385]. I visited him

Figure 92.   Paul Lauterbur (1929–2007)[386,387]. (Public domain.)

about or before then and saw his prototype. He reciprocated, when I was at NIH. At that time, there was much road construction near the Washington, DC airport. I took a wrong turn, ended up trapped on a road above ground under repair, and he missed his plane. Eventually we got off, and I left him, chagrinned, at a hotel.

He never told his wife of my fiasco, who wrote much else about me in her book on Paul after his Nobel Prize and death[390]. It's curious to see how other people view you.

Unbeknownst to me then, and perhaps to him, Paul had also gone on to make a significant contribution to the origin of life literature, on the possible enzyme-like role of amorphous surfaces[391–394], which I detailed in[73]. Robert Hazen, who has characterized the coevolution of minerals and life[395], told me that he is working on the amorphous minerals present during the origin of life, so maybe Paul's idea can be tested.

21. Synchronization of Growing Cellular Arrays (1974)[396]

Again, I tried to make new models for Conrad Waddington's then new paper on snail patterns[398]. In[396] I am acknowledged but not offered coauthorship: "The authors are grateful to Dr. Richard Gordon, who has drawn their attention to the applicability of Lindenmayer models to shell patterns." This one-way behavior was starting to wear. For instance, I wrote the code for the first version of the CT program SNARK[399], but never got a cent when it was made commercial. I thought of these snail patterns as left behind by interacting 1D pigment cells at the growing edge, using Lindenmayer methods[400] (Figure 94). I got to meet theoretical botanist Aristed Lindenmeyer (1925–1989)[401] at his home one month

Figure 93.   A cone shell[397]. (With permission under the terms of the GNU Free Documentation License, Version 1.2.) Note that the top view indicates a disruption which affected part of the pattern.

Figure 94.   3D Lindenmeyer patterns of plants. (Public domain.)

before his death. Stephen Wolfram parlayed his somewhat similar cellular automata models into a theory of the universe[402]. I collected shells from shops around the world, that had experienced damage to the growing edge (as in Figure 93), usually worse), then recovered, with the pigment patterns having to resume. Whether or not this modelling is realistic, and not just pleasing to the eye, might therefore be testable. This project needs a turntable and camera to "unroll" each snail's pattern.

22. Three-dimensional Reconstruction from Projections: A Review of Algorithms (1974)[403]

Gabor Herman (Figure 68) and I wrote a then comprehensive survey of CT algorithms, including some history relating them to classical tomography methods back to the 1920s[403]. This was reproduced in a conference publication on request from the organizer after my talk[404], this time without my resistance.

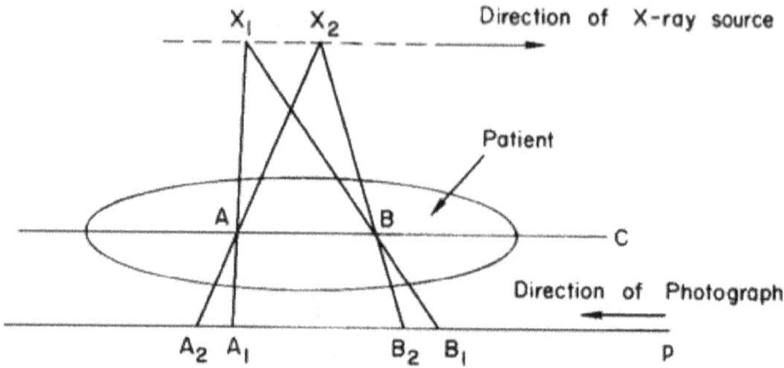

Figure 95.   One form of classical tomography[403]. The plane C is always projected onto the same part of the x-ray film, while planes above and below C are blurred out. (With permission of Elsevier.)

23. Digest of Technical Papers, Topical Meeting on Image Processing for 2-D and 3-D Reconstruction from Projections: Theory and Practice in Medicine and the Physical Sciences (1975)[405,406]

At this meeting I introduced my ARTIST (ART Intended for Storage Tubes) algorithm for positron tomography[408,409]. It was what became to be known as a "greedy algorithm"[410], but nevertheless produced a plausible image at greatly reduced dose. It was ignored by those working on PET (Positron Emission Tomography) even though I presented it later at a PET conference[411]. Later work with graduate student Barbara Pawlak introduced density estimation for PET[412]. Cf.[413].

Robert Bender (Figure 48) and I organized a course on reconstruction from projections[414], shortly after the EMI scanner was announced. It was a dud: no one signed up. We struck too early. I had learned in the interim that radio astronomy used much the same methods, and thus organized a meeting with the help of radio astronomer Ron Bracewell[415,416] (Figure 97) at Stanford

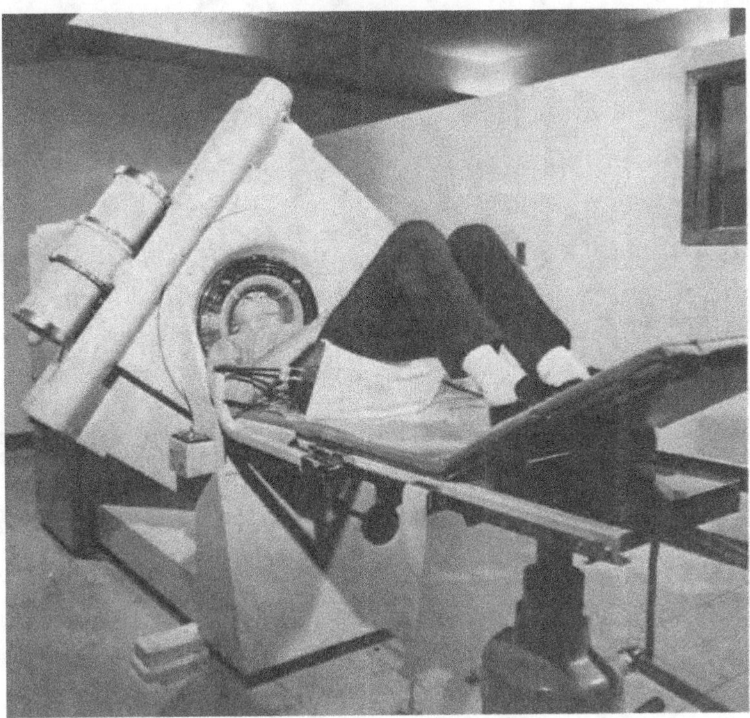

Figure 96.   The EMI scanner for the brain[407], UC. One irony is that the initial images were no better than axial tomography of the brain from the 1940s in Japan in a book I've lost track of.

University, who contributed to medical CT[417–419]. I made one mistake, rejecting a paper as irrelevant, and thus cutting off a colleague I respected, Hans Bremmerman, whose work on trans-computationalism[420] I was well aware of, and probably led to *On Monte Carlo Algebra*[252]. Later I reviewed a quite relevant book by his student Myron Katz[421], so both should have been there. The meeting nevertheless went well and covered much of what we knew then. I included a microfiche bibliography of about 100 references[422] which nowadays would be over ½ million. At that time before Internet, radio astronomers were still working near their radio telescopes, and thus began my world tour of radio telescope sites, at Greenbank, West Virginia, Arecebo at Puerto Rico (where I took the gondola to walk near the antenna, Figure 21), and Westerbroek, Netherlands[423]. I even made a small contribution towards radio astronomy in Canada[424]. In Arecebo I discussed with radio astronomer James Walker (Figure 20) the possibility of using fossilized cyanobacteria, available for instance in the Gunflint chert, to measure the spacing of heterocysts as a measure of pre-historic atmospheric nitrogen. This work was taken up by Sanjoy Som much later[64–66,425–427].

While at Stanford, I met Doug Boyd (Figure 98), who later ran a company building a fast, all-electronic CT[428,429], the Imatron, in which he imaged me and my father's hearts. This cine-CT scanner used a steered electron beam so fast that it could image a beating heart without blurring, many times a second, which seemed the future of computed tomography. The electron beam was so intense that corrections for the space charge, repulsion between electrons, had to be made. Unfortunately, he sold to a large CT company, which buried the idea. While in Austin, Texas, I visited

Figure 97.   Ron Bracewell (1921–2007), (UC.)

Figure 98.   Doug Boyd[431]. (UC.)

a small company that built a CT with photon counting detectors. These experiences expanded my awareness of what might be possible[430]. I also visited Philips in the Netherlands and consulted with lawyers in Chicago who were defending Johnson & Johnson from a lawsuit by EMI for patent infringement. Although I concluded that EMI had patented the long-known mathematics of linear algebra, we were disappointed when Johnson & Johnson folded. I learned that cooperation with companies was a one-way affair, a confusing reality that I never understood. Few pursue the future.

I met Geoffrey Hounsfield (1919–2004)[432], the inventor of the EMI scanner, to get access to the raw data from his brain scanner at George Washington University, where I was an Adjunct Professor of Radiology while working at NIH. Hounsfield was mathematically naïve, unaware that one could actually solve underdetermined equations. When I showed him how to correct an error in the raw data (the centroids of the parallel projections met at a circle, rather than a single point), I never heard from him again. This refusal to acknowledge errors by EMI, such as missing whole 3D sectors of the brain, may be what led to collapse of their leading market share. In general, CT scanner manufacturers seemed to take algorithms from published papers, made it difficult to get raw data from their machines, and acted as though our public algorithms were proprietary to them. I met with people from one pediatric CT scanner, who asked who in the hell was I to suggest how they might reduce the dose to kids, even though I had published the first paper on dose reduction in CT[433]. I eventually gave up.

The users, radiologists, also demanded sharp pictures, which required high doses (estimated at $16 \times$ the dose of a standard X-ray image) and disregard of the effect of this on their patients. (This attitude seemed prevalent: One of my kids had ordinary X-rays taken, and a radiologist across the street wanted to redo them himself, rather than requesting those films.) This may have led to the demise of an early General Electric CT with deliberately reduced dose. I wrote the first paper on dose reduction in CT[433] (cf.[434]), when I was at NIH. Despite derisive audience laughter when I suggested it might take a whole week of computer time per patient[435,436] with computer speed at the time (some physics computations took months), we issued a contract, which, being then a government employee, I couldn't apply for. Ironically, Gabor Herman got it.

Richard Webber (Figure 99), who also then worked at NIH, inspired my interest in dental radiology[437,438] and steered x-ray microbeams[439].

A curious aspect of this is that I became aware of not just doing research for good fun, but also to do some good.

In any case, Hounsfield got the Nobel Prize for the EMI scanner, while the committee ignored the work by a group of mathematicians who actually designed the first CT scanner a

Figure 99. Richard Webber[440]. (Public domain.)

Figure 100. Malvin Kalos> (With permission of Smithsonian Institution Archives, Accession 90–105, Science Service Records, Image No. SIA2008-4550.)

decade earlier, but were not helped by clinicians when they asked[441] (Figure 100), except for the later theoretical work by Allan Cormack[442–445] (Figure 128), who shared the Nobel Prize with Hounsfield[446].

24. Image Reconstruction from Projections (1975)[407]

Boris Vainshtein (Figure 102) showed that by smearing a projection across a blank film, from many angles, an approximate image could be reconstructed[414] (Figure 101). Reconstruction from projections did much better. I visited his lab in Moscow[447], where he proudly showed me his computer, which filled a room, while I had at home a desktop computer more powerful.

This[407] was my first article in *Scientific American*, which at the time was still based in the USA before moving to Germany as part of Springer Nature, without changing its name. It was a popularized introduction to reconstruction from projections, showing the relationship to moiré patterns. It was also the last paper I wrote with Gabor Herman, after a deliberate decision to work on my own. I was sent the master for the cover (Figure 103), a computer colorized cross section of someone's abdomen. I left it for a

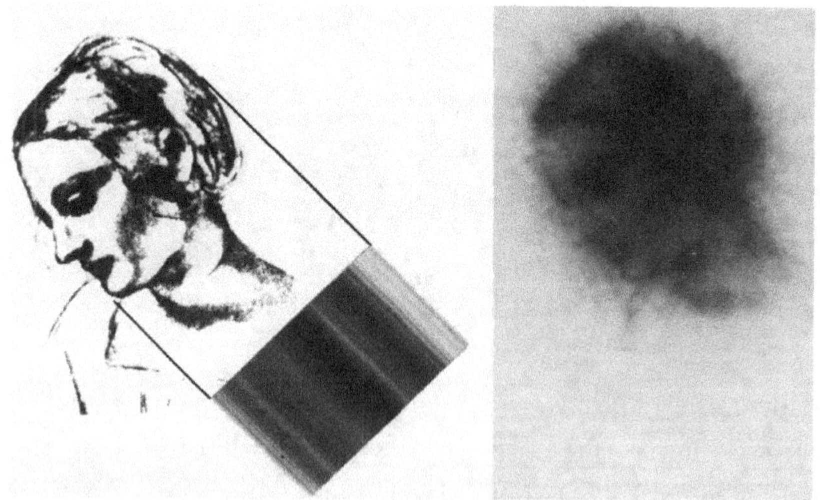

Figure 101. From[407]. (UC.)

Figure 102. Boris Vainshtein (1921–1996)[448]. (With permission of the International Union of Crystallography.)

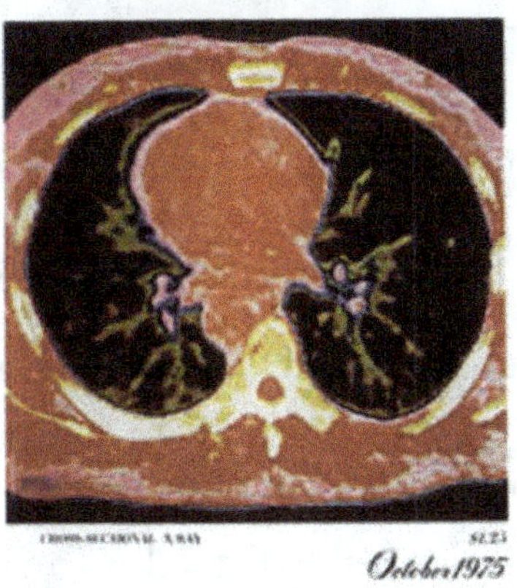

October 1975

Figure 103. *Scientific American* cover[407]. (UC.)

while in the CT room at George Washington University, and it was stolen. A compliment? That was Diana Gordon's (Figure 10) attitude when one of her paintings was stolen at an outdoor art show in Chicago, which I attended.

25. PROC10 — An Image Processing Program for the PDP-10: Operation and Description (1976)[449]

While at NIH, I transferred from Mathematical Biology in Arthritis to Lewis Lipkin's Image Processing in Cancer[450]. This[449] is a manual to which I contributed. We had an intercom. One day we heard Peter Lemkin[451] say "If you put your peanuts on top of your [computer] terminal, they get nice and warm", but all heard a near homonym for "peanuts". At one point someone who had no scientific competency was brought in to run the lab, with Lewis still there. I rebelled and was moved to the administrative building, until I left for the University of Manitoba. Alfred (Chris) Coulombre[452,453] (Figure 107), whose NIH embryology course I had attended, wrote 100 letters for me for job openings. I got a few interviews, but no job. That came after I met some people from Winnipeg at a meeting in Pittsburgh and was invited to apply there. I had to look up Winnipeg after I got the phone call, never having heard of it. I had arrived in February, 1978, deep winter. As an example of getting used to Canada's winter in my new parka, the water in the ground floor toilet froze while the movers worked. To my surprise, the items moved included a garbage can full of my Maryland garbage.

Thus, I was rescued from my administrative isolation. Lipkin's group went into exile in a lab off the main NIH campus. On retirement, Cris gave up his endeavors in embryology (eye morphogenesis in chicks), an example that puzzled me, but which I did not follow. Much later, I attended a few meetings of the University of Manitoba Retirement Society, all of whom seemed to have given up academic pursuits.

Others from whom I learned image processing were Azriel Rosenfeld (Figure 105) and Judith Prewitt (Figure 106).

While with Lewis (Figure 104), I introduced 3D deconvolution via ART for light microscopy[456], perhaps well before others discovered it[457].

Figure 104.   Lewis E. Lipkin (1925–2021)[450]. (With permission of Peter Lemkin.)

Figure 105.   Azriel Rosenfeld (1931–2004)[454]. (Under the Creative Commons Attribution-Share Alike 4.0 International license.)

Figure 106.   Judith Prewitt[455]. (Public domain.)

Figure 107.   Cris Coloumbre (1922–2012)[452]. (Public domain.)

26. Halftone Graphics on Computer Terminals with Storage Display Tubes (1976)[458]

I had been introduced to computer graphics at a Vancouver meeting, and decided to never go back. (My host lived in Bellingham, Washington, USA, and never got stopped by a border guard while commuting to Vancouver, Canada until accompanied by a young, bearded man: me.) The first electronic computer displays worked like the toy Etch A Sketch: dark or light, could not be erased except by erasing the whole screen[459]. But I needed halftones for images, so, working with two medical students at George Washington University, Lee Silver and Darrell Rigel (who later became a dermatologist, Figure 108), we selected characters to overprint to attain grey scale images. Not earthshaking, but given that my first computer image with Robert Bender (Figure 48) was overprinted on computer printouts on large, continuous form paper, spread out on the floor in a conference room, with him standing on a desk to get some distance, with me walking over the image with a marking pen at the end of a stick, the storage tube display was quite an advancement. Similarly, getting images into a computer was difficult, costing $1000/month to rent a drum densitometer[460], for instance, flatbed scanners being prohibitive ($250,000).

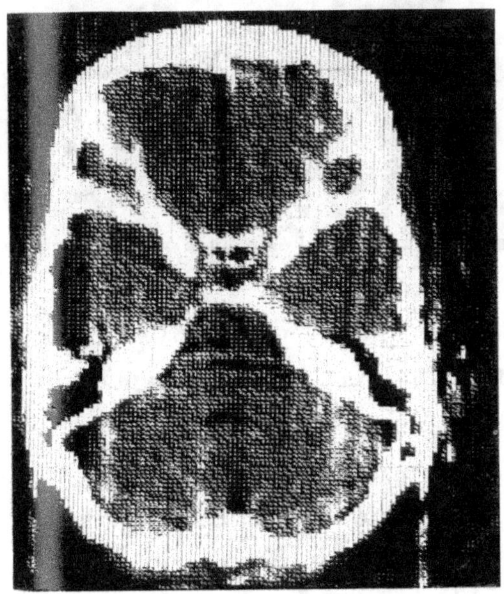

Figure 108. A brain CT scan illustrated by overprinting on a storage tube display[458].

Figure 109.   Darrell Rigel. (With his permission.)

Figure 110.   Richard Feldmann. (With his permission.)

Digital cameras were $50,000 in the 1970s. Scanners, much better, now cost about $100, sometimes less. Digital cameras are now in every smart phone. We pushed what we had, or could afford. My proposal at NIH to provide a scanner service was met with derision and a reminder of my low status there ("Expert").

One high point at the NIH computer center was friendship with Richard Feldmann (Figure110), who played with plastic toy models, which I did subsequently, and was best known for his pioneering 3D computer models of proteins[461]. We met frequently in the NIH computer center. He set an excellent example of freedom of thought, and visited us in Manitoba at our rough cabin.

27. Changes in the Shape of the Developing Vertebrate Nervous System Analyzed Experimentally, Mathematically and by Computer Simulation (1976)[462]

I worked many years with Antone Jacobson[464] (Figure 112), an embryologist at the University of Texas, Austin. He had room length shelves of sectioned salamander embryos (California newt, *Taricha torosa*), and some time-lapse movies, which we had to image by using a single frame movie projector and sketch the projections on paper. These were the days before we had digitized movies (videos). The computer simulations, which I did at NIH, took so long that I had to run them overnight on a PDP10. Some debugging was done at home with a teletype terminal. One night the PDP10 computer started asking me questions that I unfortunately answered. The result was that everyone's work from the day before was deleted. The next night, to my wry amusement, only my day's work vanished.

It is retrospectively irritating to see my ideas attributed to him, though of course his experimental observations provided the data:

"In subsequent years, Antone and Richard Gordon would use the data generated from this and similar studies to develop a computational model of neural morphogenesis. In a landmark paper in *The Journal of Experimental Zoology* in 1976, Antone put forth his notion

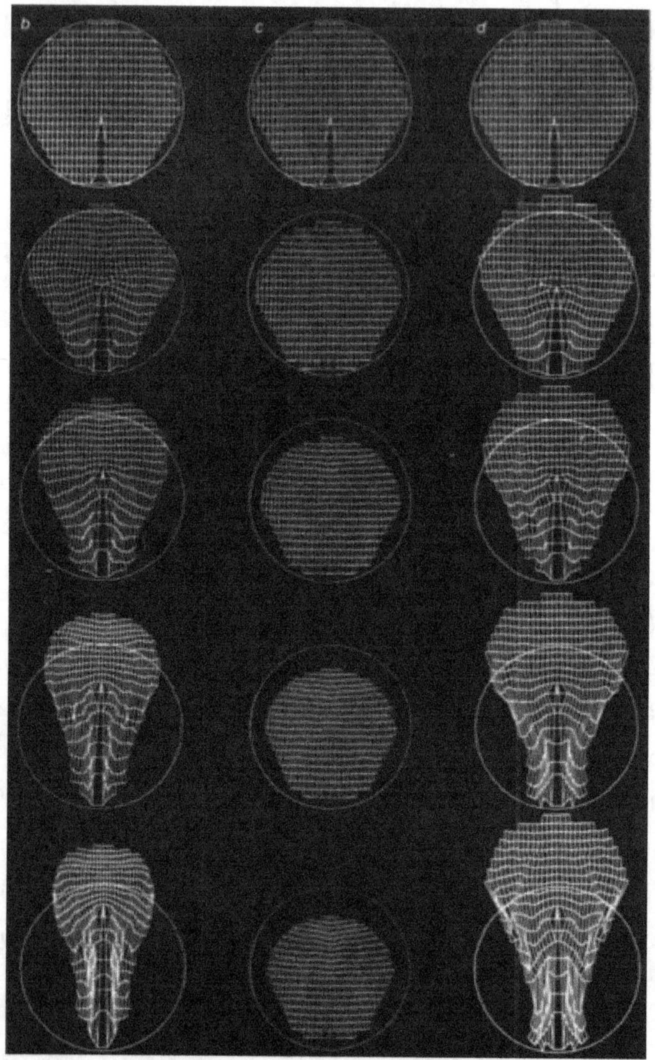

Figure 111. The left column is my computer simulation implementing represen-
tations of both the notochord and our empirical observations of the shrinkage of the
apical surfaces of the neural plate cells. The second column is a simulation without
the notochord. The third column shows the misshapen oversized neural plate if the
shrinkage is left out of the simulation. Thus, both are needed for normal development.
(From[463], UC.)

that morphogenesis research should proceed as a closed loop of experiment, formal mathematics, and computer modeling, whereby modeling generated predictions that could be experimentally validated and also inform further experiments in order to improve the model)[462]. With over 50 figures (and an appendix that is remarkable even with today's insatiable appetite for supplemental data), this paper coined the term "morphodynamics" and laid out what remains the standard blueprint today for studies of embryonic morphogenesis"[464].

They didn't ask me how we worked so well together and complemented one another. This reminds me of the caption to the cartoon in[201]: "To be honest, I would have never invented the wheel if not for Urg's ground breaking theoretical work with the circle". Let this be a warning to all who might want to call themselves theoretical biologists. Just because "Urg" is a play on ur = original, doesn't mean that Urg got some credit.

Figure 112.   Antone Jacobson (1929–2017)[465]. (UC.)

We had the beginnings of AI then, in the form of a "doctor" program that asked you questions about your health. In those days, you basically lived in the computer center. At NIH I observed people getting addicted to the game Moon Lander, unsuccessfully tried it a few times (success meant your stick avatar did not crash and got to eat a hamburger from MacDonald's on the Moon), and swore off computer games along with tobacco, alcohol and drugs for the rest of my life. I prefer the high of ordinary consciousness, though I did do a bit of reading on "altered states of consciousness" (never indulged, only getting as far as buying a package of morning glory seeds, but not ingesting them) and had a "mature" graduate student, Tony De Luca, interested in consciousness[466].

To Antone's chagrine, after[462], when *Scientific American* had us do an article[463], I became first author, as he had insisted on first authorship on the earlier *Journal of Experimental Zoology* article[462]. As I recall, my later ideas on embryology were deemed by *Scientific American* as "politically too dangerous" for them. We also published[467–469]. At that time, Sweden was my ideal, and I was jealous that Antone got to go there and present our work without me[470].

Two articles about me, one without mentioning my name, did appear in *Scientific American*[471,472].

28. Vision Begins with Direct Reconstruction of the Retinal Image, How the Brain Sees and Stores Pictures (1977)[473]

Helmut Hirsch (Figure 17) was my best friend in high school, who had a career in vision research. I followed much of the work in vision, for instance meeting Karl Pribram (1919)[474] at his Stanford lab, who considered the brain as a hologram[475]. I suggested that the elongated visual receptive fields acted much like rays in CT (Figure 113), and was able to thereby simulate the visual discriminations of kittens that Helmut raised with restricted visual experience. The paper was published in a festschrift for his father, and followed up by a University of Manitoba medical student[476].

I learned a bit about color vision and its genetic variations[149]. Thus, when I had a stroke in 2020 and was recovering in the hospital,

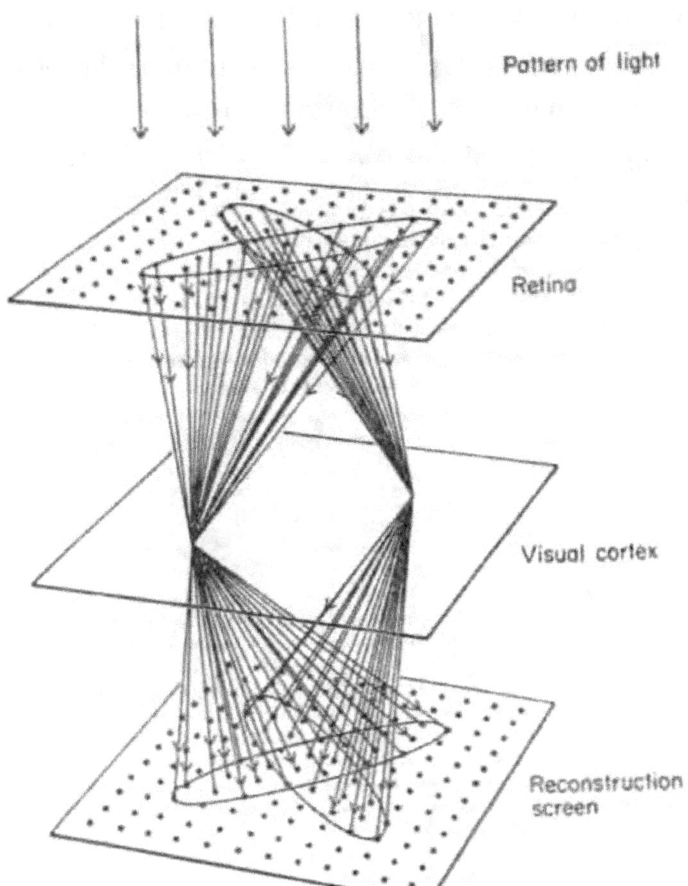

Figure 113. A crude CT model for the role of receptive fields in visual perception[473].

I was amazed to see the blue fabric on a visitor's chair briefly turn red. The room was unchanged. Current researchers in color vision I contacted had no explanation for this object specific hallucination. I don't recommend my way of triggering this hallucination, so can't recommend any way of approaching solving it.

29. High-speed Reconstruction of the Finest Details Available in X-ray Projections (1977)[477]

This was my first publication aimed at breast cancer. I did run length encoding of each projection and then backprojected the starting and ending points of each run across the 2D area for the reconstruction. Instead of using pixels, this approach divided the reconstruction area into polygons. These were to be filled in

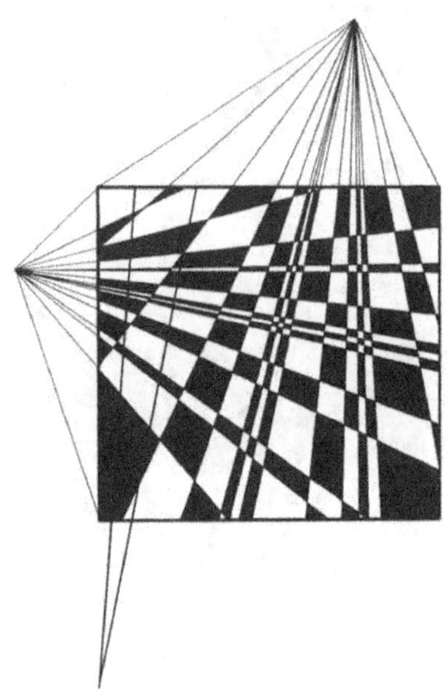

Figure 114.   Division of the plane via run length encoding of three projections[477], own drawing. (UC.)

with a reconstruction algorithm such as ART. Afterwards, I realized that there would be many tiny polygons that did not reflect structure, and abandoned this approach.

Triangular pixels have been used for cross-bore tomography in geology[478], which usually has an overall geometry similar to breast compression paddles. Thus, our work on the RBYRCT algorithms for breast cancer may help in underground imaging.

30. Computer Department for Health Sciences (CDHS), University of Manitoba (1978–1981)

I was hired by the University of Manitoba as an Associate Professor to administer this computer center, which was getting a new computer, with academic appointments in Pathology and Radiology, as my choice. The previous Director of CDHS had died about 3 years previously. The CDHS was run by one of its staff in the interim, who rebelled perhaps due to his *de facto* demotion upon my arrival. I had to fire him to get the new computer installed, late, as a result of the need for a false floor to cover the array of electric cables. It was my first and last episode as an administrator. I had to deal with a staff of 20 of all ages and abilities, decreasing budget each year, old fashioned demands from faculty such as insistence that the new computer include a card reader, giving other faculty members "permission" to purchase their own computer, etc. I inherited one of the first computer-controlled microscopes of 1960s vintage and established the Quantitative Morphology Unit, which no one used. I felt my research productivity decline. Thus, at the end of my 3-year term I gave the standard two weeks' notice. I recall that the CDHS lasted another year under another staff member. There were some faculty jealous of the CDHS budget.

During my stint in CDHS, I hired a pregnant statistician against the directive of the Dean. Two other female statisticians filled in for her during her pregnancy leave, which went smoothly.

I learned a lot, but not how to manage other people. On leaving the CDHS I had my computer tapes transferred to hard disks, which ran up a bill of $36,000, which I hadn't anticipated and couldn't afford. They settled for a small fraction.

31. The Shaping of Tissues in Embryos (1978)[463]

This was our work[462] boiled down for *Scientific American*, with it's origins in DÁrcy Thompson's transformations acknowledged (Figure 115).

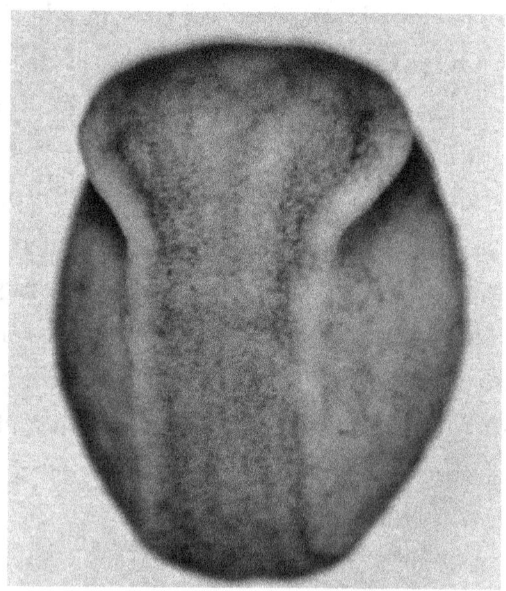

Figure 115.   The neural plate in a salamander embryo, *Ambystoma mexicanum*. The variegated pigmentation makes it possible to track individual cells[463]. (UC.)

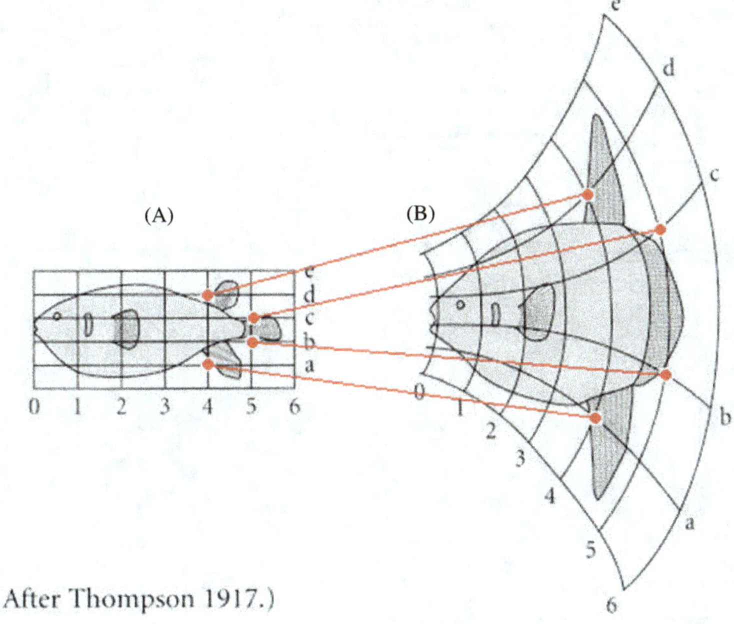

(After Thompson 1917.)

Figure 116. A D'Arcy Thompson transformation, (Public domain.)

32. Computed Tomography for Rural Manitoba via Teleradiology (1979)[479]

Given my involvement with underdeveloped countries, I proposed a CT design that was hand operated, with the projection data sent by phone to a medical computer center[479,480]. We designed a system with up to 4 telephones and acoustic couplers, before Internet, to send the images with error correction, during which I met with the chief of Norway House, Manitoba, who brought up its wider ramifications. At the time, we had excellent cooperation and help from an executive, Dennis Wardop (Figure 118) at the Manitoba Telephone System (MT)[481]. We actually installed a system from Brandon to Winnipeg, Manitoba, but the radiologists in Brandon

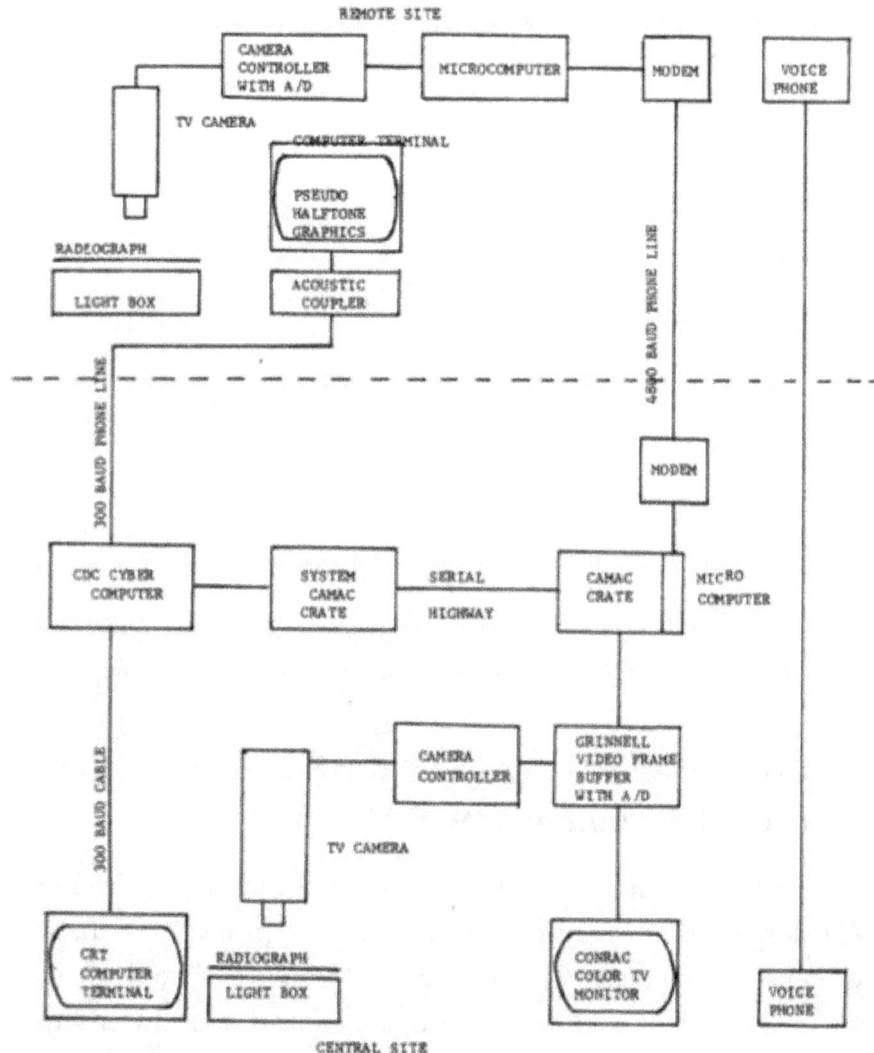

Figure 117. This is what it took then to transmit radiographs or multiple CT images before the Internet[480].

were not interested in advice from their radiology colleagues in Winnipeg, so it was never used for patients.

At the time, Canada had generated 50% of the worldwide literature on teleradiology, but not one manufactured system. I made

Figure 118.   Dennis Wardrop (1932–2013)[482]. (UC.)

many academic and other attempts to overcome this disparity, but had no business smarts[437,479–481,483–487], and thus failed. Canada now has old teleradiology equipment imported from the US, replaceable by the Internet.

Chapter 3
1980s

33. The Need for Cross-fertilization between the Fields of Profile Inversion and Computed Tomography (1981)[488]

This was a foray into satellite imagery of Earth and trying to establish a relationship to CT. CT later figured in 3D imaging of our atmosphere[489,490].

34. Rotating Microscope for "LANDSAT" Photography of Vertebrate Embryos (1982)[491]

LANDSAT was the first satellite that looked down at the Earth while photographing it. As noted above, I attended a meeting on such imaging[492]. I had an apparatus built that would carry a stereomicroscope around an embryo, trying to image it from all directions[491,493]. Wayne Brodland (Figure 65) designed its vibration isolation platform. The rotating microscope never worked. A flaw in the computer running it sent the computer back to its manufacturer, who inadvertently erased all the data we had collected. The axolotl embryo was placed on the top end of a capillary tube narrower than the embryo, with a slight negative pressure to hold it, which caused the obviously viscoelastic embryo to "drip" down the tube. But the idea of imaging the whole surface of an axolotl embryo persists to this day, with a "flipping microscope"[494] that Natalie and

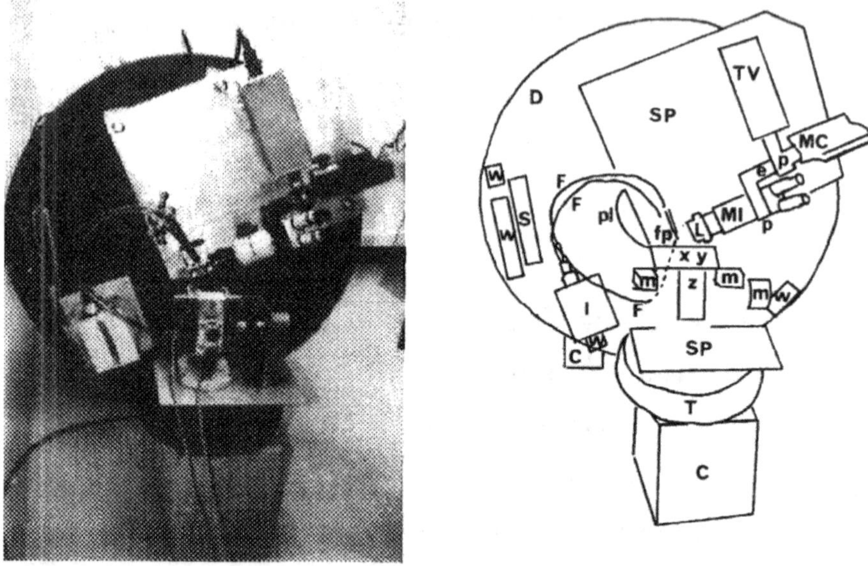

Figure 119.   The rotating microscope before being mounted on its vibration isolation platform[491], with author's copyright.

I showed feasible during a weeklong visit to the hospitable laboratory of Randal Voss, University of Kentucky, and then various designs built by my former student, Susan Crawford-Young, engineer and artist[494] (Figure 186). With the help of Marco Gruwel, we explored micro-MRI of axolotl embryos, which penetrates their optical opacity[495]. However, the strong magnetic field of 11.7 Tesla altered their development from an early stage, rendering the images worthless for understanding normal embryo development.

35. The Future of Computers in Medicine (1983)[496]

It is amusing how little one can look into the future, and I'm no exception. For example, I didn't see past computer tapes for storage, which at most held 140 megabytes on a 10.5" reel. The 3D

Figure 120.   Eluemuno Blyden[498]. (With his permission.)

movie (not mine) that I showed while seeking a Veterinarian professorship at the University of Guelph of a simple stiff face took 45 minutes per frame to generate. Now we do much more complicated 3D imagery for video games at 30 frames per second and I own a 4 terrabyte SSD that fits in the palm of my hand. Eluemuno Blyden[497] (Figure 120) and I glanced into the future of medicine at a later date.

36. Random Number Generators for Microcomputers (1983)[499]

Much of my work used stochastic methods. There were two methods then for generating random numbers: 1) a radioactive source; 2) pseudo-random numbers. The cycle for the latter before the numbers generated repeated depended on the word size of the computer used, which was especially small then for microcomputers. This was an attempt to overcome the word size limit, done with electron microscopists.

Xiohua Zhou (Figure 60) and I made a small contribution to random binary pictures[500].

37. Methionine Enkephalin-induced Changes in Pigmentation of Zebrafish (Cyprinidae, *Brachydanio rerio*) and Related Species and Varieties, Measured Videodensitometrically. I. Zebrafish (1983)[501]

Zebrafish of various strains provided patterns of numerous sorts, and I incompetently tried raising them with a homemade fancy UV water sterilization apparatus while I was at NIH. But they wouldn't breed for me. The problem turned out to be that the UV device heated the copper tubing near it, driving copper into the water, sterilizing the fish. It was yet another foray into being an experimentalist that ended badly. I was a charter member of the Pigment Cell

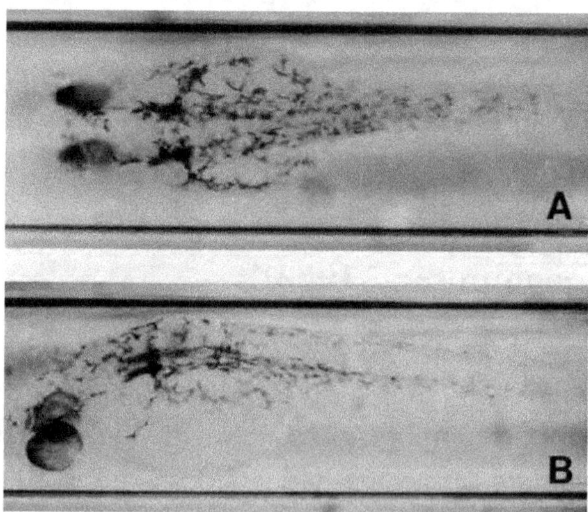

Figure 121.   Two views of a zebrafish embryo, rotated in a capillary tube, which permits its optical computed tomography. Note early stripes and yolk sac pigment cells. From[502] with permission of John Wiley and Sons.

Society. For a while I attended cat shows, feeling like I was walking in Darwin's footsteps when he attended fancy pigeon shows, hoping that cat lovers might donate pelts of their deceased show pets, but none did. I was particularly interested in the American shorthair tabby, and its crosses with striped cats, both of which had bull's eye patterns on their sides. How did pigment cells presumably migrate from the skin over the backbone to the flanks to form such patterns? This puzzle remains unanswered. I also visited a dalmatian breeder, who culled puppies not meeting their standards, keeping them frozen until they could be interred, to try to gain insights into spot patterns. Variegated plants exhibit similar puzzles. I visited a lab focussing on them and occasionally purchased some. I photographed pelts of zebras and giraffes in backrooms at the Field Museum of Natural History in Chicago and live at the Buffalo Zoo. I collected books on their variants in patterning.

A medical student, Martin Mummery, did time-lapse microscopy of the pigment cells over the zebrafish embryo sac, which moved like a sparsely connected sheet[503], but the paper was rejected and never improved, as he was a summer student, back in medical school and not further available. I developed a method for getting multiple views of a copper-free zebrafish embryo for computed tomography of these transparent embryos[502,504], but didn't follow through. The work just got reported at a conference[505].

Another medical student, Shurjeel Choudhri, followed up the sac pigment cells with electron microscopy[506]. He is now Senior Vice-President and Head of Medical and Scientific Affairs, Bayer Canada.

Although I supervised 50 undergraduate students, usually in summers, their short efforts, however intense and successful, only twice resulted in published papers.

Sophie Levina, a pathologist and refusnik from the USSR, joined me in Winnipeg and wrote this fish endocrinology paper[501] and another[507]. But I never got into the endocrinology involved. I had attended a small conference of scientist refusniks in Moscow, USSR and was followed by a police car. To get some perspective, I visited a group of East African students in Moscow, aimed there somehow by French speaking African students, despite my lack of French.

In dermatology, which I learned about via my interest in pigment cells, if one has a suspect nevus, the dermatologist cuts it off with a resection margin, and sends it to a pathologist to diagnose it and make sure all cancerous cells have been removed. This became my model for search and destroy of small breast tumors. Thus, the incomplete work with pigment cells has proven valuable. My awareness of the medical problem also led to nevoscopy, described later.

I tried, unsuccessfully, to get a collaborator to work on imaging low contrast nevi in dark skinned people.

38. Computational Embryology of the Vertebrate Nervous System (1983)[456]

This was a summary of the methods at that time. It also introduced the idea of deconvoluting the 3D point spread function for microscopy using a computed tomography approach. I was postdocing with Lewis Lipkin (Figure 104) at the time, but I was not enough of a practicing Jew to fit in.

39. Expanding the Dynamic Range of X-ray Videodensitometry Using Ordinary Image Digitizing Devices (1984)[508]

During my career I impatiently saw imaging equipment come and go when replaced with cheaper, better devices. But it's hard to

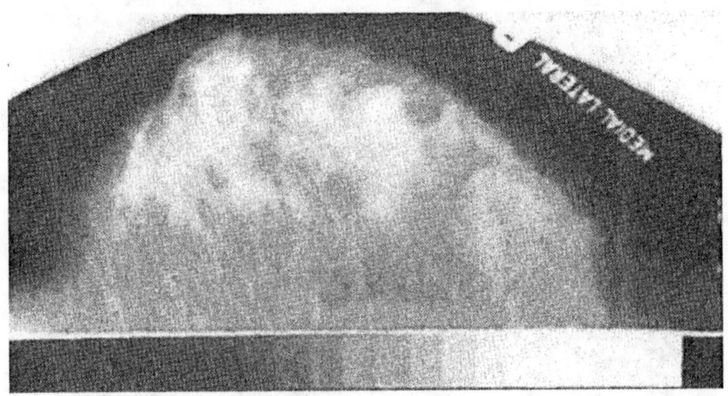

Figure 122. An expanded range mammogram with a density scale appended[508]. (With permission of Optica.)

wait. This paper is an example, where the limited optical density range of video cameras (replaced by densitometers, then scanners) was pushed to correspond to the much higher range of X-ray film.

40. Nevoscopy: Three-dimensional Computed Tomography for Nevi and Melanomas In situ by Transillumination (1984)[509]

I recall waking up excited one night, to look at a nevus ("beauty spot") on my arm with the tube of my home microscope and a flashlight held nearby into my skin to back illuminate it. It became obvious that the 3D structure of the nevus was observable (at least for light colored skin)[510]. Atam Dhawan (Figure 124), who had taken an image processing course I designed with my post-doc Rangaraj Rangayyan (Figure 125), joined me as a graduate student. This was then the ultimate challenge for computed tomography: we designed an attachment to my stereomicroscope that allowed photography of a nevus from just three directions, straight down 0°, 45°, and 90°, simultaneously. (90° required that the skin was pressed, so that the nevus rose.) We called it

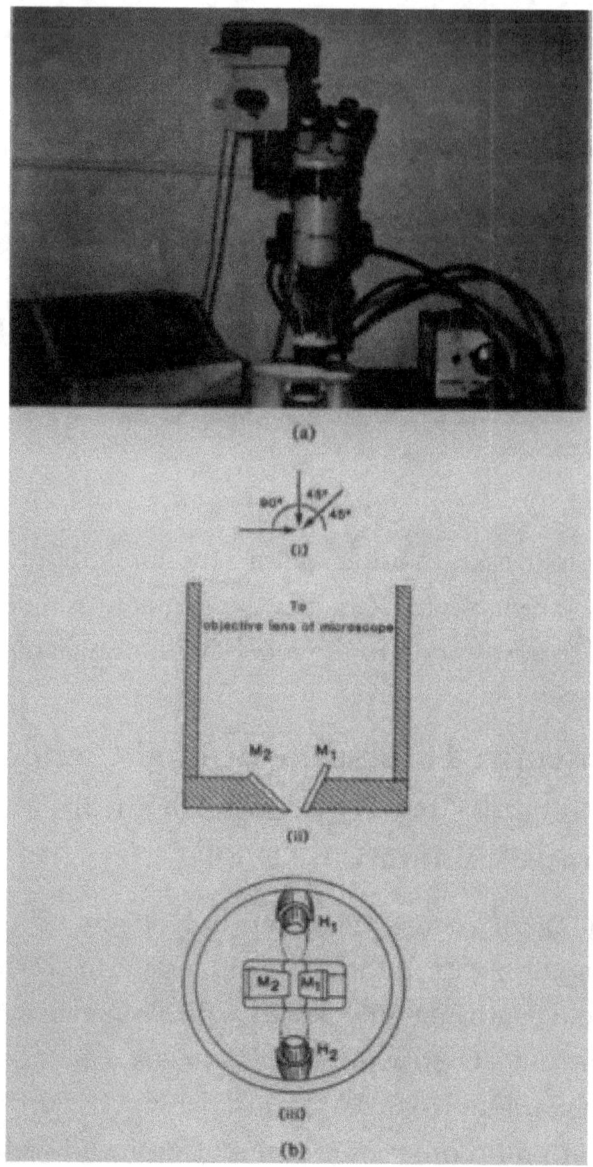

Figure 123. The original nevoscope, with a diagram of its construction. Two optical fibers H_1 and H_2 provided the light into the skin[509]. (With permission of the IEEE.)

Figure 124.   Atam Dhawan. (From[524] UC.)

Figure 125.   Rangaraj Rangayan. (With his permission.)

the "nevoscope". Now amongst dermatologists, it was well known then that the depth of a nevus was the best prognostic characteristic. Poor Atam, an engineering student, then imaged a nevus on a cadaver in Anatomy, had it cut out and sectioned, and while our 3-view reconstruction was poor, it nevertheless gave the depth of the nevus. A number of papers followed[509,511–523].

The point spread function (PSF) of ART obviously varied across an image, so I visited mathematician James Nagy, who was using Toeplitz matrices to try a spatially nonuniform deconvolution for the Hubble telescope. (Someone had placed a lens backwards[525], but later this was fixed by a manned space flight to the telescope). But we never tried Toeplitz matrices for limited view tomography, using a "simpler" spatially uniform Weiner deconvolution[516,526] (cf.[527]). It nevertheless took Atam 6 months to get it to work. Despite our approximation that the point spread function was uniform, the result was a spectacular improvement over ART with limited views[517,526]. A small literature on deconvolution of limited view CTs has appeared since, which Rangaraj and student Paul Soble contributed to[527–537]. Since then machine learning may be doing as well or better[538].

Nariman Sepehri, an expert on robotics, and I hired a summer student from France, to try to carry a camera over the whole body, looking for nevi. Instead, the student spent all his time pursuing a girlfriend in Winnipeg. So, while there has been much progress in diagnosing images of nevi without a nevoscope, the whole-body approach remains untackled. Melanoma, if not detected early, before it metastasizes, has a 95% kill rate. Early detection leads to 95% survival. Some of us took an unsuccessful crack at crowd funding[539].

Atam went on as a professor in the Department of Electrical & Computer Engineering, New Jersey Institute of Technology, to improve nevoscopy, but it never caught on.

41. Combining Multiple-imaging Techniques for In vivo Pathology: A Quantitative Method for Coupling New Imaging Modalities (1984)[540]

Jack Coumans of Philips used two projectors and two screens to show these two slides simultaneously (Figure 126). I immediately recognized them as 2D projections from a 4D space which could

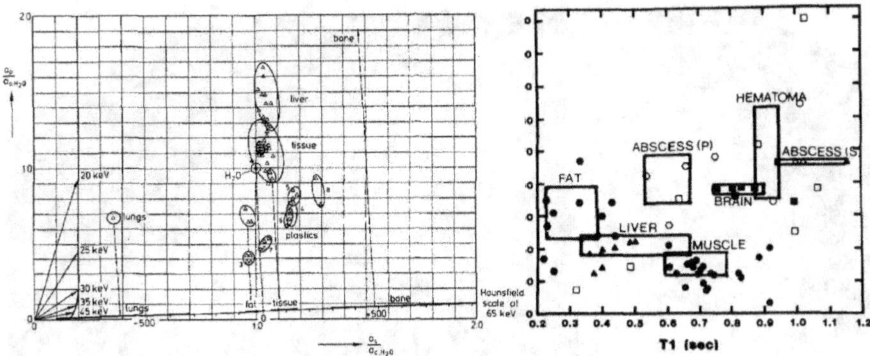

Figure 126. Two MRI measurements and two x-ray measurements, 2D projections from 4D[540] . (With permission of John Wiley and Sons.)

contain more information than the images considered separately, due to possible correlations. Thus, we proposed that such data could be combined in 4D[540], allowing a richer analysis when multiple imaging techniques are used on the same patient, because of possible correlations between the imaging modalities. No one in medicine has followed up this higher dimensional approach, despite many combined imaging machines, such as PET/CT (some listed in[201]). I guess it's hard to think in 4D.

42. Feature Enhancement of Film Mammograms Using using fixed and adaptive neighborhoods (1984)

For a brief period, xeromammography was widely touted. Xerography depends on the accumulation of powdered black pigment due to electrons imaged on a plate. As such, it is tied to the forces between electrons and the pigment granules, and therefore has a particular physical distance over which it works and enhances the contrast of the primary image. This may or may not match the distance between detector pixels. Therefore, to get maximum contrast, the pixel size might have to be increased. Most imaging is

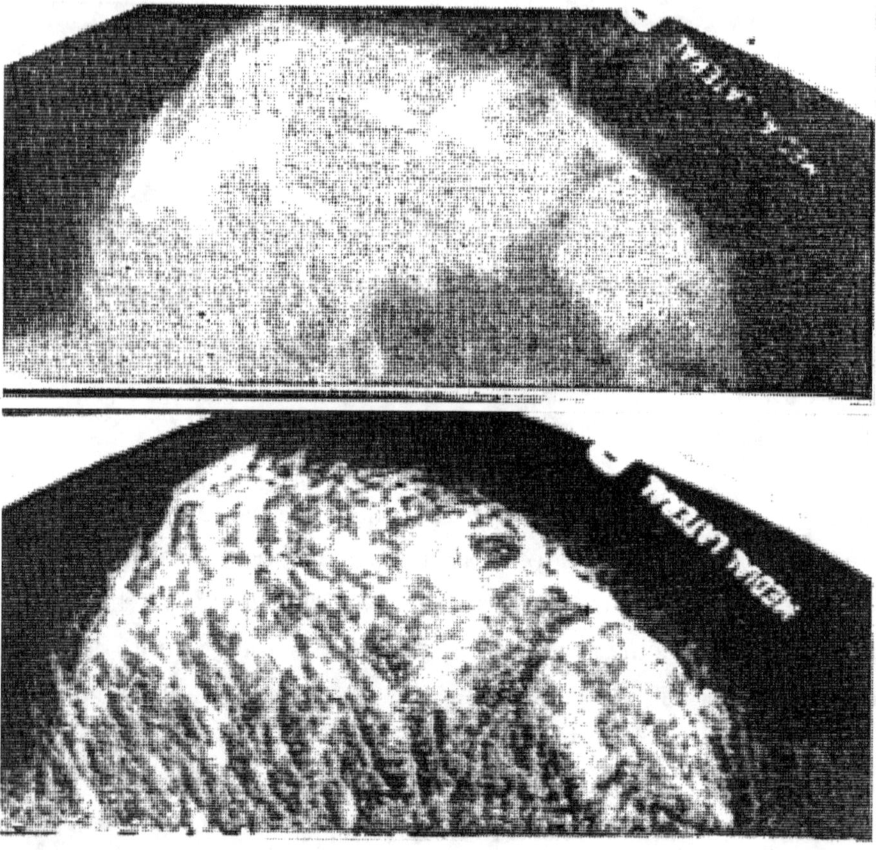

Figure 127.   A mammogram and the same one processed by adaptive neighborhood contrast enhancement[541]. (With permission of Optica Publishing Group.)

done with square arrays of detector pixels. To create larger pixels they can be grouped: 1×1, 3×3, 5×5, etc. Odd numbers are chosen, because then there is always a central, physical pixel. If one then plots the contrast of the other pixels in the adaptive neighborhood against the middle pixels, this function may increase for a while and then taper off. Its first maximum is taken as the center pixel's contrast. Thus, the pixel size adapts to the local structure in the image and enhances it. Rangaraj Rangayyan (Figure 125) and I proposed this as an alternative to xeromammography[542], starting from digitized ordinary film mammograms, thus obtained with no further dose.

While xeromammography faded away, Rangaraj went on at the University of Calgary to develop more sophisticated methods of adaptive neighborhood image processing, and a number of other people took it up.

Actually, our first paper using adaptive neighborhoods was for a problem that plagued medical CT: if the person had a metal implant, it blocked the X-rays entirely, and with ordinary CT algorithms caused streaks across the image, making diagnosis difficult. We proposed *Streak preventive image reconstruction with ART and adaptive filtering*[543] (cf.[543,544]) that alleviated the problem.

We wrote a bit more[520,521,545] and got some flack[523] about an image entropy measure we used, to which we replied[522].

43. Algorithms for Limited-view Computed Tomography: An Annotated Bibliography and a Challenge (1985)[518]

Two cases came up in many fields besides tilt stage electron microscopy: limited number of views in computed tomography and limited number of rays per view. We included the projection data for an image in this paper, without the image, attempting to challenge the reader to get the best reconstruction they could and identify the image. No one took up the challenge. It remains.

Suppose a given object is radiation sensitive, allowing a maximum of p photons. If one could arrange irradiation with $q < p$ photons per m views, what is the optimal value of q, with $mq = p$? As far as I know, this has not been addressed for cone or fan beam CT, and is an open question. Obviously, when q is small, statistics become important. Can one go to the limit of $q = 1$?

Rangaraj Rangayyan and I took forays into limited view CT[528,529,546].

44. Industrial Applications of Computed Tomography and NMR imaging: An OSA Topical Meeting (1985)[547,548]

I ran one more meeting for the Optical Society of America. Allan M. Cormack[549,550] (Figure 128), Nobel Prize winner, was treated like a normal participant, except that I scheduled his talk to the end, so no one would leave early. This was probably the first conference on industrial applications of CT.

45. Toward Robotic X-ray Vision: New Directions for Computed Tomography (1985)[552]

In this review, I introduced my heuristic "proof" of Kennan Smith's Indeterminacy Theorem[553], in my attempt to understand it (Figure 129). I also, with some humor, suggested a CT scanner

Figure 128.   Allen Cormack (1924–1998)[551]. (Licensed under the Creative Commons Attribution 3.0 Germany license.)

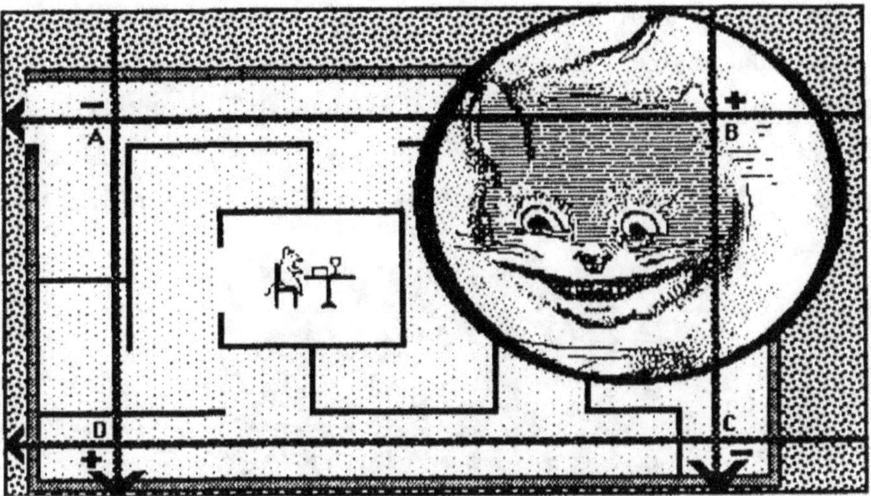

Figure 129. The + and – indicate how it is possible to change a CT point by point without altering the projections[552]. (With permission of Optica.)

based on a ring of sources and detectors placed horizontally, that worked so fast that one could drop a wriggling baby through the ring and get a whole-body scan as it fell though the ring. An uncracked raw egg represented the baby.

Kennan T. Smith was a pure mathematician who got involved[554–557] when a neurologist friend of his asked how it was possible that a patient who he was sure had a brain tumor yet it didn't show up on a CT. He visited me for about a week, inspiring a book on CT emphasizing his approach, but the real analysis math proved to be over my head, so it was never completed.

46. Display of 3D Anisotropic Images from Limited-view Computed Tomograms (1985)[558]

Before there were movies, there were peep boxes or slots through which one could view a succession of images, giving the impression of movement, going back to Roman times, while riding chariots. I realized that even just two views were adequate to give an impression of depth, using stereoscopy and the peephole concept, an idea that was tackled by engineering undergraduate Kirby Jaman. Afterwards I experimented with 3D visualization of with varying densities of dots in 3D visualized from two directions. With Ira Rai (Figure 130), Jenny Wagner and Vandana Vinayak (Figure 131), we

Figure 130.   Ira Rai, astrophysics student. (With her permission.)

Figure 131.   Vandana Vinayak. (With her permission.)

are now trying to determine to what extent two views through the gravitational lensing of a galaxy cluster can envision the 3D structure of a more distant galaxy seen through a pair of galactic gravitational lenses, and reviewing 3D imaging of galaxies.

47. From Statistical Mechanics to Molecular Biology: A Festschrift for Terrell L. Hill (1987)[92]

This special issue honoring my PhD mentor, Terrell Hill (Figure 26), extended to two issues[92,559]. He was a shy person, and did not want to attend a symposium in his honor, so we settled on this Festschrift. He bought many copies for his relatives.

48. The Cytoskeletal Mechanics of Brain Morphogenesis. Cell State Splitters Cause Primary Neural Induction (1987)[560]

I had asked Wayne Brodland (Figure 65), then a graduate student in Mechanical Engineering, University of Manitoba, to design a vibration isolation platform for the rotating microscope. He told me later that he almost became my student. Thus, when he became a postdoc at the University of Waterloo, Ontario, I joined him on my 6-month sabbatical to write about the mechanics of cell state splitters and the waves they generate[560,561] (cf.[378]). His supervisor's hobby farm near Waterloo felt claustrophobic compared to the farms of Manitoba, measured in sections (squares ¼ mile on a side) while his German guests expounded on its vastness. Spinoffs included the first mechanical theory of the role of intermediate filaments in supporting microtubules, making the microtubules much less vulnerable to compression and buckling[562–564], the first mechanical explanation of their role in the "integrity" of a cell. (This awareness of buckling was later applied to diatoms[565–567].) We explained, with a simple model, why inverted axolotl eggs sometimes developed normally, and sometimes didn't[568]. Russ Flint was a high school physics teacher who spent his sabbatical year in my lab on this work.

We drifted apart when Wayne was advised to remove my students from an already submitted paper because "it had too many authors". This was resolved a year later by his own administration when it became clear that an embryology group at the University of Waterloo he worked with was also left out[569]. All involved became coauthors, but he had destroyed a fruitful collaboration. It was now possible to track cells in embryos by computer[570]. Wayne went on to produce a number of independent papers on the mechanics of embryonic tissues. In particular, he generalized a method developed by medical student Murray Stein and me[571] (Figure 132) to estimate the forces between cells in epithelia[572]. Murray became a psychiatrist and wrote[573] with an acknowledgement to me and others "for kindling my interest in research".

Figure 132.   Murray Stein, (UC.)

49. Oblique Sampling of Projections for Direct Three-dimensional Reconstruction (1987)[574]

By using a ray width that depends on angle, the raysums could vary smoothly along each projection. Fixed ray widths, dictated by today's detectors, make the raysums a jagged function.

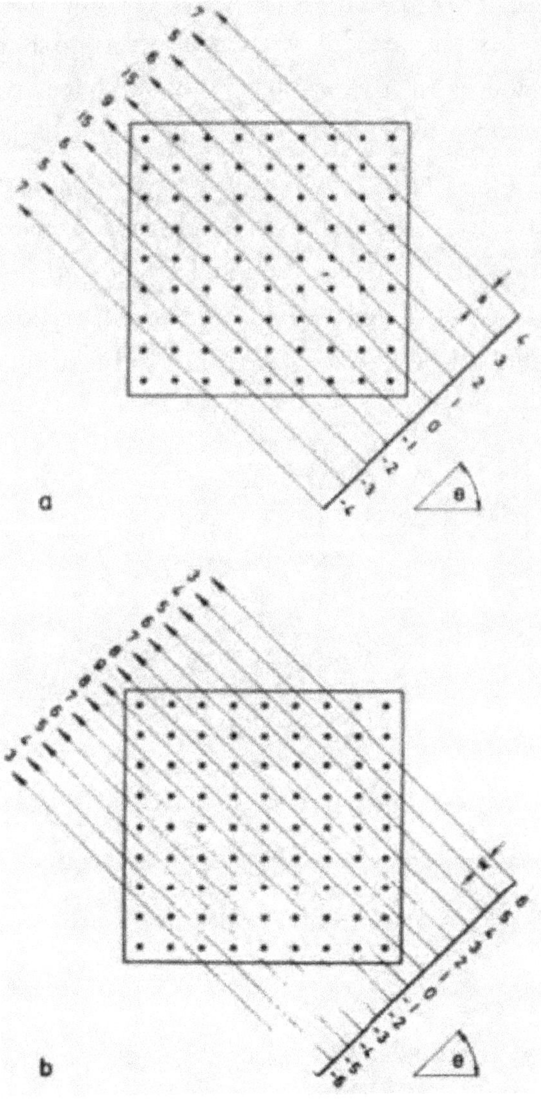

Figure 133.   Ray width varying with angle[574]. (With permission of Elsevier.)

50. A Retaliatory Role for Algal Projectiles, with Implications for the Mechanochemistry of Diatom Gliding Motility (1987)[575]

One day, in a crowded public bus on my way to coteach a small 4th year Botany class in phycology with Phil Isaac at the University of Manitoba, I jotted down notes for what became my lecture and subsequently this paper. This was the only time I made significant use of calculus in my career, a subject that is so often used as an antiquated gatekeeper for students. The single-celled alga *Ochromonas tuberculatus* makes a cell-sized jump when disturbed. As the barb (discobolocyst) sticks out, I calculated that it has the speed of a rifle bullet, which I suggested penetrates the mouth of any organism, such as *Daphnia*, trying to devour it. Thus, the alga "retaliates". The energetics had implications for diatom

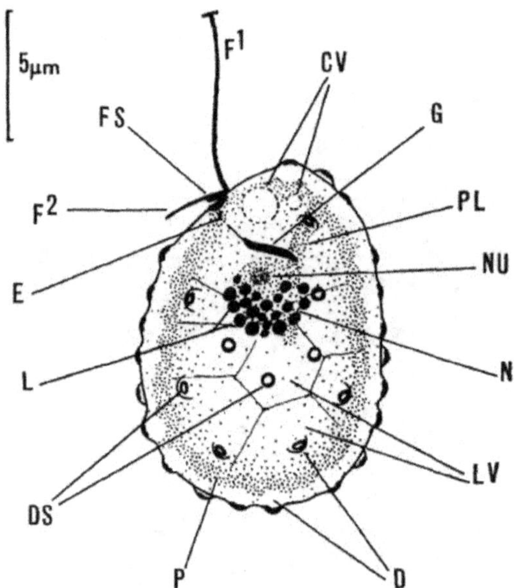

Figure 134. *Ochromonas tuberculatus.* (With permission of Cambridge University Press.)

motility. Many other living things go "pop", making for an entertaining lecture[576,577].

51. Neurulation (1989)[578]

This chapter was a review of axolotl neurulation. See also[579].

52. A Critical Review of the Physics and Statistics of Condoms and Their Role in Individual versus Societal Survival of the AIDS epidemic (1989)[30]

The world was being attacked by HIV/AIDS and I wanted to contribute. An undergraduate and I decided to tackle the problem of spread in small networks of people, but I was no expert on networks, so we got turned down. I wrote a newspaper oped suggesting that *There is no such thing as safe sex*[580] (cf.[581]). On the basis of my newspaper article[580], I was invited to an andrology conference in Vancouver, where, after my talk, I was shunned by the medical participants, apparently because they could no longer just prescribe condoms for their patients to prevent AIDS. In reaction to this weird behavior, I delved into the efficacy of condoms from every angle, producing *A critical review of the physics and statistics of condoms and their role in individual versus societal survival of the AIDS epidemic*[30]. The editor put a caveat at the beginning of the paper, reproduced in the Preface to this book. A scientist at the then reputable US Center for Disease Control read the paper and invited me to an AIDS conference in Switzerland. I had a poster demonstrating that most leakage from a condom occurred through the big hole at the top[582], based on red dye experiments similar to those I saw in a dentistry seminar. This upset an obviously gay

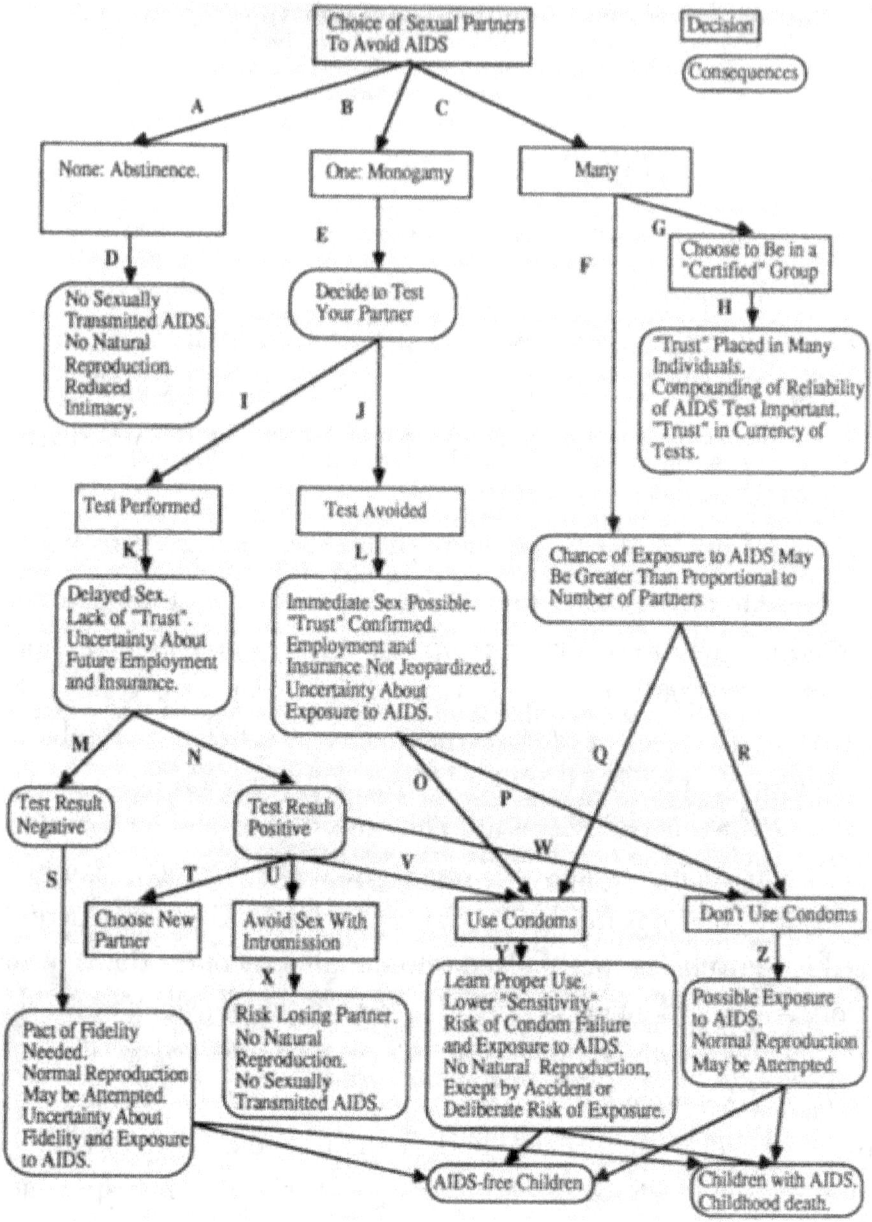

Figure 135. AIDS decision web[30] before treatment was available. (With permission of Taylor & Francis Informa UK Ltd — Journals.)

organizer of the conference, who told me that my poster was anti-gay, a thought that had never crossed my mind. The next day there was a banner across my poster declaring that it was not an official contribution to the conference. My abstract was dropped from the printed proceedings.

On getting in an airplane to leave Switzerland, I found myself seated next to the delegates from Thailand. I told them my conclusions: 1) condoms do not provide guaranteed protection for the individual; 2) but if everyone would wear them, the R_0 estimate indicated that the epidemic would halt. Apparently, shortly afterwards, Thailand started the 100% Condom Program[583], which saved millions of lives. Natalie would like to believe I was the cause. I'm not acknowledged.

I read[584] for perspective on denial in the gay community.

This led to new designs for better condoms, which we attempted to commercialize unsuccessfully, and were visited by Swedish condom inventor Axel Rappe[585] (Figure 136) with whom we cooperated for a while.

Figure 136. Axel G:Son Rappe. (With his permission.)

I got involved with Abba Gumel (Figure 137), a mathematical epidemiologist, and we, together with others, modelled the spread of HIV/AIDS slowed by condoms[586]. With John Tyson[587] (Figure 138), a faculty gynecologist, whom I "knew" originally

Figure 137.   Abba Gumel[590]. (Under the Creative Commons Attribution-Share Alike 4.0 International license.)

Figure 138.   John Tyson (1935–2022)[587], (UC.)

Figure 139. Bob McLeod[593], (UC.)

from his television show in Maryland, I previously generalized the problem and showed that statistical runs of fatal diseases could lead to extinction of a species[588], perhaps an important aspect of evolution (cf.[589]). Most species have lasted only 3 million years. As with phase transitions in small systems, I was getting more awareness of the importance of rare events. This figured later in my approach to the origin of life[73]. John drove a Cadillac and was disturbed by my orange Datsun with rust spots painted with bright orange metal paint polka dots. We nevertheless got along.

Probably through his participation in this paper[586], I met Bob McLeod (Figure 139) and learned a bit about mining data, even in real time, from the Internet[591]. We also wrote about hospital infections[592].

We tried to inform Bill Gates on how he could contribute to halting worldwide HIV/AIDS, when he spoke at huge AIDS conference in Toronto[594], but doubt that his gatekeepers ever told him of our efforts[595–599]. No one from the Gates Foundation attended, even remotely.

Chapter 4
1990s

53. Inexpensive Computed Tomography for Remote Areas (1990)[600]

I did a lot of thinking about how to make computed tomography work in developing countries, in outer space, and in rural areas[480]. While none of these pre-Internet thoughts (film based with television camera images sent via teleradiology) got past the ideas stage, they certainly colored my future work on computed tomography.

54. Defects due to Compression of Amphibian Embryos at Primary Neural Induction (1991)[601]

Although this is just an abstract, it is an important loose end. The use of mechanical forces to alter embryonic development might lead to important insights. The work of Lev Beloussov (Figure 140) on mechanically creating multiple neural plates[602] is similar in spirit. Our paths crossed in the USA[603,604] and twice in Russia, and we wrote some papers together[375,605–609].

The old concept of a morphogenetic field[610] may be instantiated as the trajectory of a differentiation wave. Beloussov sought to relate them to biophoton emission[611]. An amusing sidelight is that

Figure 140.   Lev Beloussov (1935–2017)[614]. (With permission of Abir Igamberdiev.)

Natalie and I were invited to a small meeting organized by physicists interested in learning some biology[612]: in the introductory session, upon learning that we were embryologists, one invitee who planned to speak on Rupert Sheldrake's[613] bizarre concept of morphogenetic fields and their ability to be in intercontinental "morphic resonance" immediately left. We were asked to explain the basis for his behavior, despite feeling like people with emphysema due to the altitude in Telluride, in the mountains of Colorado, to which we hadn't yet acclimated.

During the trip our car conked out, and was repaired by a former helicopter repairwoman, which much impressed young Lana (Figure 10). As we had told her that Barbies had no brains, she found a pilot Barbie doll "with brains" and we got her a plastic helicopter big enough to hold the doll. The trip down the Rocky mountains was scary, as a slip off the winding road meant 1000s of feet drop down. On a walk I grabbed Lana as she almost fell off a cliff.

55. On Lifting the Inherent Limitations of Positron Emission Tomography by Using Magnetic Fields (MagPET) (1992)[615]

In positron emission tomography (PET), the location of the positron is not known until it is annihilated, producing a pair of detectable gamma rays that move in opposite directions. Dan Rickey (Figure 141), then a Physics undergraduate, showed that attainable magnetic fields would confine the trajectory of a positron, thus increasing resolution. He is now an imaging physicist with his own pet axolotl. This proposal has been briefly followed up[617], but seems to have yet had no impact on PET scanners.

While a magnetic field will only confine positrons in two dimensions, it would be interesting to know how fast it could be rotated without harming patients, to also get better 3D resolution.

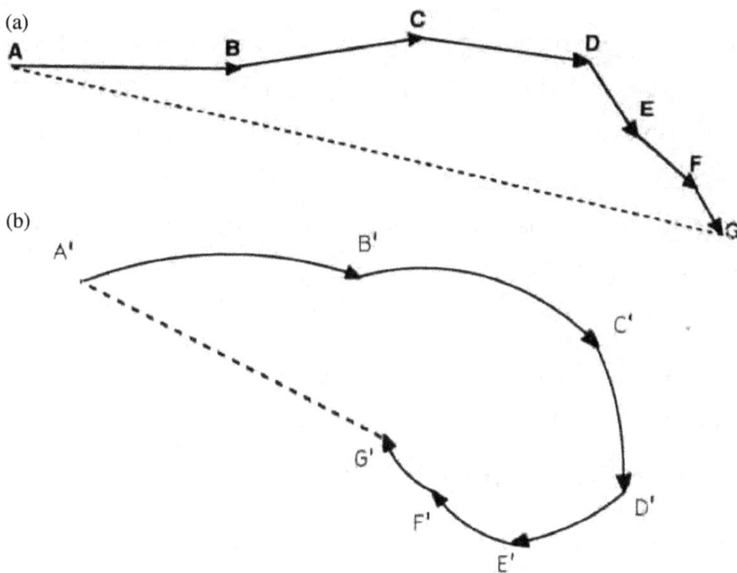

Figure 141. In a magnetic field, a positron moves less total distance before annihilating with an electron[615]. (Journal *Automedica* has ceased to exist, UC.)

Figure 142. Daniel Rickey[616]. (With his permission.)

56. Soliton Waves on the Vertebrate Embryo, an Excitable Medium (1992)[618]

Solitons are peculiar waves that annihilate if two of the same amplitude meet, rather than passing through one another, as sine waves (ripples) in water do. Here, we proposed that differentiation waves in embryos are solitons. Note that differentiation waves are also kink waves, as they change the tissue through which they propagate to a new tissue.

57. Contrast Enhancement Using "Feature Pixels" or "Fixels" for Pixel-independent Image Processing (1992)[619]

Graduate student Huaiqun Guan (Figure 143), now a medical physicist, worked on a number of improvements to computed tomography algorithms. This one is an aspect of adaptive neighborhoods:

"By introducing the concept of feature pixels, or fixels, we present a new method to perform the contrast enhancement based

Figure 143.   Huaiqun Guan[620]. (UC.)

on the features of images. Fixels are defined as contiguous and nonoverlapping regions of an image, each representing a local feature of the image. All pixels in a fixel have similar gray levels. We will show different ways to partition an image into fixels. Examples of medical images that have been enhanced by using their decomposition into fixels are presented under the optimal definition of a fixel and optimal choice of the enhancement function. Fixels are an improvement over the concept of pixels and thus form a new starting point for all image-processing algorithms"[619].

Huaiqun went on to optimize ART with respect to the order in which projections are processed[621–623]. I also expanded the fixel idea to a pointillism point of view for image processing and computed tomography[201].

58. Differentiation Waves in *Drosophila* and the Axolotl (1993)[624]

The morphogenetic furrow of the eye imaginal disk of the fruit-fly *Drosophila melanogaster* was seen by us as an example of a differentiation wave. We thus examined it in quantitative detail, after Diana Gordon (Figure 10) hand-fixed the poster in[626], by closing the boundary around each cell, tedious work that took her 10 hours (Figure 144).

In retrospect, David Suzuki (Figure 145) had actually beaten us on insect eyes as differentiation waves. He had shown that in the temperature sensitive *shibire* mutant of *Drosophila*, one could use the developing eye as a time-lapse record for a wave of determination that crossed the eye imaginal disc[628]. But he didn't tie

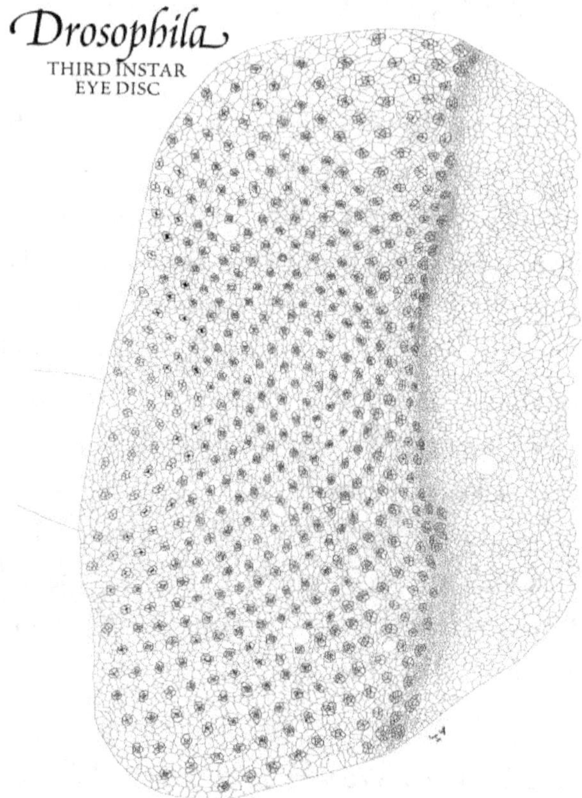

Figure 144.   A camera obscura sketch of all of the cells in a *Drosophila* eye imaginal disk, touched up extensively by Diana Gordon[625] to make each cell closed. (With permission of Elsevier per https://www.stm-assoc.org/intellectual-property/permissions/permissions-guidelines/.)

Figure 145.   David Suzuki[627]. Uunder the Creative Commons Attribution-Share Alike 3.0 Unported license.)

this seminal observation to differentiation in general, as we have done, based on our work on axolotl embryos. Instead, he went on to produce nature documentaries for television and became an environmental catastrophist[627]. Fame may be an impediment to understanding.

59. Grant Agencies versus the Search for Truth (1993)[629]

When I first became a professor in 1978, at the University of Manitoba, I applied for a grant, and received severely less money than I asked for, because "theoretical biologists don't need money", as if their students didn't have to eat. Some years later I decided to take on the Canadian grant agencies, leafleting faculty meetings, writing about their foibles, and even getting on national radio[630].

A number of papers followed, starting with the one above, including plots of the increasing ages of grant recipients over time[631]. I joined a group of Canadian scientists, who were also trying to reform the grant system, CARRF, (Canadians for Responsible Research Funding)[632–636]. I then met Bryan Poulin (Figure 162) through a friend of my wife, a Business Administration professor at Lakehead University, Ontario, Canada, who promoted ethical capitalism. We tackled the system with titles such as *Cost of the NSERC science grant peer review system exceeds the cost of giving every qualified researcher a baseline grant*[637,638] and *How to organize science funding: The new Canadian Institutes for Health Research (CIHR), an opportunity to vastly increase innovation*[639]. We gave testimony in the Canadian Parliament[640–642] and wrote about alternative granting mechanisms[643]. The net result was the impression that I was being "turkeyed" by University of Manitoba administrators and publicly berated by the then head of CIHR. I learned that grant agencies always know better than scientists what science should be done, vast bureaucracies replaceable at much lower cost by check printing presses. The anonymity of the peer review system is questionable[644,645]. One might think that, with titles such as *Give profs more money*[646–648], the faculty might support me, but grantsmanship keeps them divided.

Bryan and I branched out, writing a review of a book on grant alternatives[649], tackling the energy shortage[650], and proposing a World Minimum Wage[651].

I tried to get CARRF renamed UNFUN (The Union of Unfunded Scientists). Key CARRF members passed away, and so did CARRF. Now that I'm retired, I can't qualify for grants, and leave the issue to coming generations.

60. Nuclear State Splitting: A Working Model for the Mechanochemical Coupling of Differentiation "Waves" with the Controlling Genes (Master Genes) (1993)[339,652]

I met Natalie a few times before beginning a relationship that eventually led to our respective third marriages. Our initial meeting was when I substituted for a faculty member at a biology undergraduate lab, discussing paper chromatography and embryology for hours, until closing time. Natalie checked on my motivation with one of my female students and joined my lab a few months later. She started in my lab by excerpting fragments of epithelia tissue from axolotl embryos, and spotted what appeared to be a

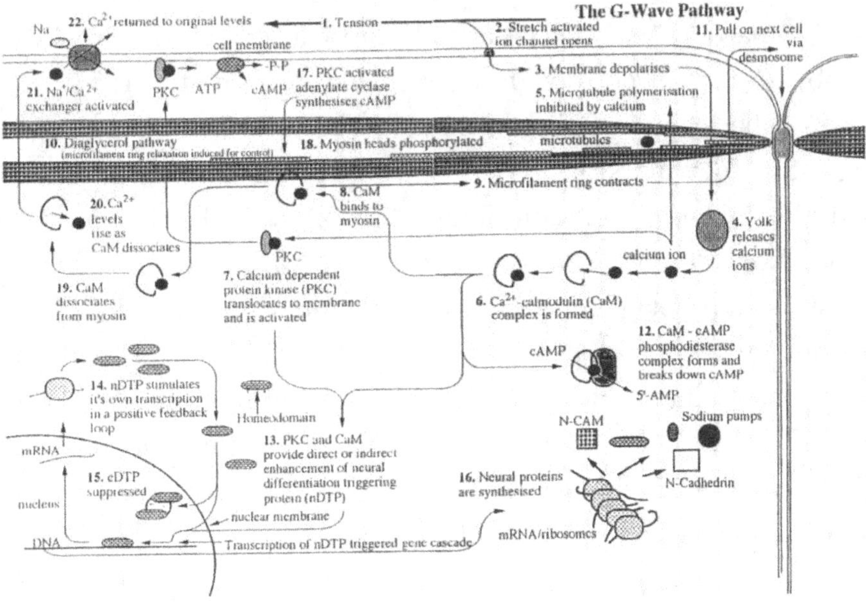

Figure 146.   This was an attempt by Natalie to estimate the biochemical pathway from the cell state splitter to the nucleus[339,652].

wave of contraction going over a fragment. This was confirmed in intact embryos, leading to a novel theory of cell differentiation in embryos. We tried to get a grant to do the clinching molecular biology in tissue isolated before and after the contraction wave that leaves the neural plate in its wake. But she was then an undergraduate, and I was no molecular biologist, so we got turned down. Understanding the bias by North American molecular developmental biologists against anything but genetic explanations, we turned to a Russian journal, *Ontogenez*, which published this model (Figure 146) that Natalie eked from the literature, on how what we called the "cell state splitter"[653] at the apical end of each neural plate cell might contract or expand, sending a molecular signal to the cell's nucleus to change it into a new kind of cell (differentiate). *Ontogenez* published the article in both Russian and English[339,652]. We got married in a synagogue with classical India music provided live by my former postdoc, Rangaraj Rangayyan (Figure 125).

We have since published much, elaborating this idea, including a string of papers[333,355,569,653–659] and two books[149,183]. Involvement with Russian publishers continues occasionally[660] (cf. full article[661]). Pieter Nieuwkoop (Figure 147), who we regarded as the "dean" of classical embryology, endorsed our view in a posthumous paper we helped reconstruct[662]. One day, at the former University of Manitoba Delta Field Station (destroyed later one winter by wind driven ice off Lake Manitoba), Pieter and I went canoeing and were caught in a fierce wind. We had to abandon the canoe on the shore and walk back to the field station.

A curious side note to this is an unexplained "ghost story". Pieter Nieuwkoop, who followed and participated in our work, died at age 79 during heart surgery. A bit later I found a small hard disk on my desk, written using a word processor that my computers

Figure 147. "Pieter D. Nieuwkoop [1917–1996] with Natalie K. Björklund at Kildonan Park, Winnipeg, in 1992, discussing early axolotl embryogenesis".

could not read. Dan Rickey (Figure 141) managed to open it, and it was an incomplete manuscript that endorsed our work at its end:

> "Although a whole scala of influences seem to be at work at particular stages of development in the organisation of the CNS, the observed changes in morphogenesis may depend upon separate, successive binary decisions, as assumed by Gordon in his [cell and] nuclear state splitter hypothesis[560,652]."

To keep things above board, I contacted Pieter's colleagues in Utrecht, Netherlands, who found an identical disk. Together we did our best in completing Pieter's manuscript, just in time to get it into a special issue of the *International Journal of Developmental Biology* dedicated to embryological work in the Hubrecht Laboratory, which he had directed. While Pieter had visited us before his death, how this disk got on my desk remains a mystery. Pieter was to write the Foreword to my book[183], which was superbly done by Natalie from her detailed memory of him.

I hired medical artist Jack Butler (1937–2024)[663], who had worked at the University of Manitoba, to sculpture whole axolotl embryos to exhibit the ectoderm contraction wave in 3D. He photographed these clay models, which were printed on consecutive pages in[183] as a flip animation, preceding computer videos (what I had as a kid for homemade "movies"). It came out well[664], but the cost used up most of my royalties.

One undergraduate, Cris Martin (Figure 148), was especially productive, providing electron microscopic evidence for the cell state splitter[331,568,569,601,659,665–667]. He later became a professor at the University of Ottawa and then moved to St. George's University, Grenada, tired of the rat race for grants.

Exasperated at how we were ignored, no one coming forth to test our ideas, nor give us a grant to test them ourselves (perhaps because none of us were molecular biologists, the wall keeping people to their specialities), I tried to be explicit later with *Are we on the cusp of a new paradigm for biology? The illogic of molecular developmental biology versus Janus-faced control of embryogenesis via differentiation waves*[336]. Top-down causation goes against reductionism, so together with physicist Bashir Ahmad (Figure 149), I wrote *Is it a Janus-faced world*

Figure 148.   Cris Martin. (With his permission.)

after all? Physics is not reductionist[337] (cf. an earlier foray into reductionism[668] and [669]). However, if we're right, the proper testing will eventually be done posthumously, probably by a physicist entering biology, without the ingrained molecular developmental biology nor EvoDevo preconceptions. But then, physicists have upset biology many times.

One deduction we made is that there must be a "differentiation code" for each intermediate embryonic cell type[670]. To first order, embryonic differentiation may be described as a "differentiation tree", a notion that Bradly Alicea has been building on[332,671–675]. It is unfortunately not popular amongst the code biologists, an offshoot of the biosemioticists. I coedited a book delving into biosemiotics[676,677], but did not get deeply involved in the field. I may write a chapter speculating on the evolutionary origin of "continuing differentiation"[183].

Figure 149. Bashir Ahmad. (With his permission.)

61. Evolution Escapes Rugged Fitness Landscapes by Gene or Genome Doubling: The Blessing of Higher Dimensionality (1994)[678]

One problem with the idea of a fitness landscape for evolution is that it is often what is called rugged, with many peaks and valleys. Thus, evolution finding the peak of such a fitness landscape can be difficult in the face of many local minima that can trap it. In this paper I show how genome doubling (Figure 150), a common occurrence in evolution[679,680], allows the peak to be approached in an all-uphill manner. The idea has been followed up a bit[681,682], even though the journal I chose is not an evolution journal.

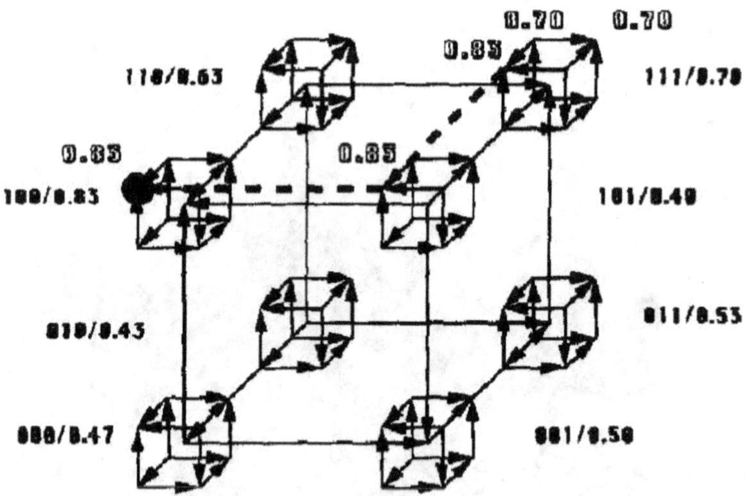

Figure 150.   An example of a genome duplication with no valleys in the fitness landscape *en route* to the peak[678]. (With permission of Elsevier Science & Technology Journals.)

62. Analysis of Pericentric Inversions (1994)[683]

Niel Finkle (Figure 151) was a medical student who worked on this project with me and Phyllis MacAlpine[684] (Figure 152), a chromosome expert in the Human Genome Project, who unfortunately died shortly after. We found some unexplained periodicities in lengths of human chromosome inversions.

Figure 151. Niel Finkle. (With permission of the Sustainable Nephrology ('SNAP') Committee requested, UC.)

Figure 152. Phyllis MacAlpine (1941–1998)[684], (UC.)

An earlier attempt by me to contribute to the Human Genome Project, with a proposal to have a subcommittee to define tissue types, was accepted, and then "rejected for political reasons"[685].

63. Mechanical Engineering of the Cytoskeleton in Developmental Biology (1994)[686]

At the time of this book I edited, I was President of the short-lived Canadian Society for Theoretical Biology[687], which died when we mistakenly merged with a larger organization, which increased

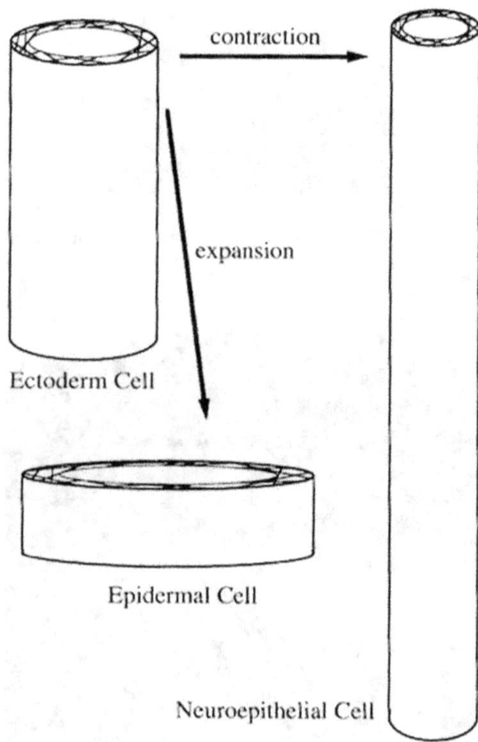

Figure 153.   The differentiation choices of embryonic ectoderm cells, with the cell state splitter schematically depicted[656]. (With permission of Academic Press/Elsevier.)

our membership fees. The late Brian Goodwin (Figure 154)[688] was co-chair for the meeting and co-editor of the book, but dropped out with "I know what you are trying to do", which might have been tied to his own views on morphogenesis[689]. I never got an explanation, but guess that I was on a different tact intellectually. Theoretical biologists don't always see eye to eye. For example, is the addition of a "sense dimension"[690] distinct from the Janus-faced viewpoint that I ended up with[335–337]? Perhaps the philosophically oriented reader will tackle this? The contributions included chapters by him[691], Lev Beloussov[692] (Figure 140), Wayne Brodland[693], Umberto De Boni[694], Don Ingber[695] and collaborations with Ryan Drum[288] (Figure 85) and Pieter Nieuwkoop[656] (Figure 147), all of whom played or would play important roles in my life.

Umberto De Boni's work on the rotation of some cell nuclei, without the cell rotating, intrigued me to the point that I put a summer student on a crude model for it[696]. The role and mechanism of nuclear rotation is still unresolved[697].

Figure 154. Brian Goodwin (1931–2009)[688] under the Creative Commons Attribution-Share Alike 3.0 Unported license.

64. Radiologic Study of Limb Regeneration in the Axolotl (1994)[698]

Martin Reed (Figure 155) is a pediatric radiologist with a keen eye for research. Axolotls readily regenerate their limbs, though sometimes not perfectly, resembling limb defects found in children. Thus, Chris Preachuk, then a Radiology resident, took up this study of axolotl limb regeneration. Ted Lyons (Figure 156), leading ultrasound in Radiology, did research on the effect of ultrasound

Figure 155.   Martin Reed[700]. (With his permission.)

Figure 156.   Ted Lyons. (With his permission.)

on human embryos that intrigued me[699], as a mechanical effect during embryogenesis.

65. Sheared Drops and Pennate Diatoms (1996)[701]

I attended a conference in France that included a talk on immiscible droplets stretched by quadripole water flow at the surface,

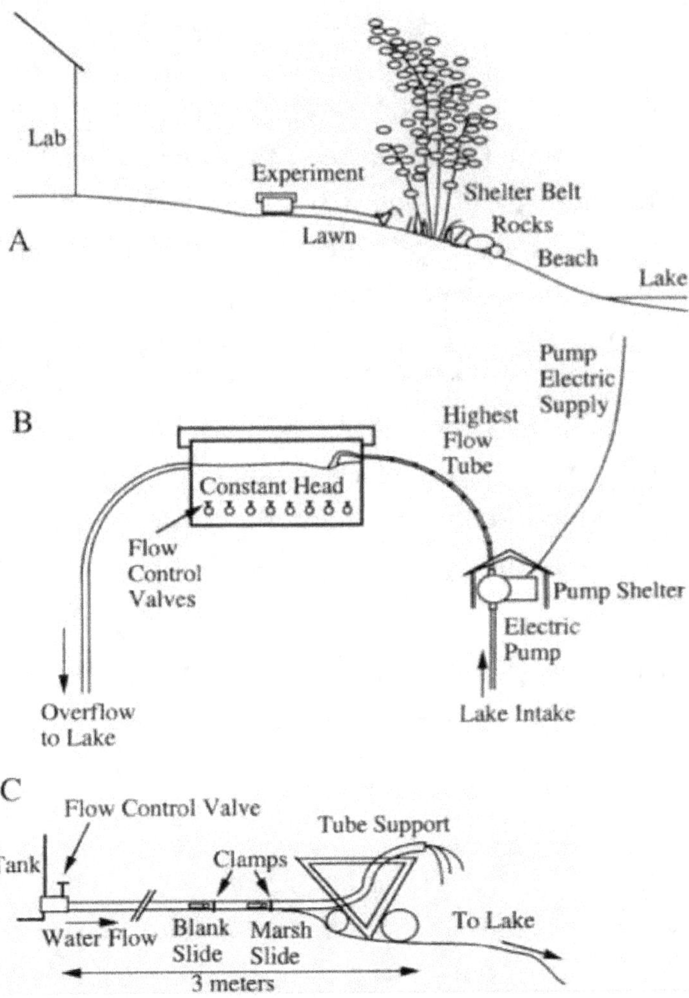

Figure 157. Experimental setup for testing diatoms under assorted flows[701]. (With permission of Schweizerbart and Borntraeger science publishers http://www.borntraeger-cramer.de/journals/nova_hedwigia.)

Figure 158.   Hedy Kling. (With her permission.)

that uncannily resembled some pennate diatoms. This led to an outdoor experiment that confirmed that long, pennate diatoms adhered to microscope slides at the highest flow rates. Thus, it was speculated that over geological time water flow could shape pennate diatoms over generations.

This work was carried out in a summer by me and Natalie at the former Delta Marsh Field Station[702,703]. Hedy Kling worked for Fisheries and Oceans Canada, adjacent to the University of Manitoba, doing scanning electron microscopy (SEM) of diatoms[701,703,704]. I collected samples of water with Lake Manitoba diatoms for her for a while.

66. On Halting Bear and Professor Hunts: Internet and the University of Manitoba Professors' Strike for Academic Freedom (1996)[705]

I was an "e-mail savvy volunteer"[705,706] in the days when e-mail was just catching on. During a faculty strike I saturated e-mail and faxes, especially in the Manitoba Government, with our message. Salient points were summarized in[707]. During the strike, I recall

Figure 159. Arnold Naimark. (With his permission.)

that the Dean of Medicine, Arnold Naimark, although on the other side, drove me once to strike headquarters. After the strike, at a ceremony honoring him, I butted in and presented him with an "academic freedom" placard from the strike. It took guts, as some said.

I spent some time during the strike in the University of Manitoba archives showing that the ratio of administrators to students was rising much faster than the ratio of professors to students.

67. Effect of 60 Hz Ambient Magnetic Fields on the Development of Axolotl Embryos (1997)[708]

Mosden Abdel-Hadi did a Master's thesis with me while working at Manitoba Hydro, the province's provider of hydroelectricity. One day we found that all of our axolotl embryos within a Helmholz coil were oriented in the same direction, but could not repeat this. It might be worth more tries. At the time, I assiduously read in the journal *Bioelectromagnetics.*

At that time, there was great concern that overhead transmission line electric fields were deleterious. My wife Natalie and I met with the rabbis of Winnipeg, where we presented the equivocal evidence of harm. They had been offered inexpensive land under transmission wires. When we suggested they could go ahead with their congregants, a somewhat genetically homogeneous group, as an experiment, they found another place for the Jewish Community Centre.

I reviewed the occasional forays over the century into electric fields in embryos in[183] and helped with a University of Calgary thesis on embryo electrorotation[709]. While there has been much exciting work in bioelectricity recently[710], it has not yet been incorporated into embryogenesis in an explanatory manner.

68. Refraction Mapping of Translucent Objects with Shack-Harman Sensor (1998)[711]

Vasyl Molebny of the Ukraine visited me in Winnipeg and proposed this method for 3D imaging of axolotl embryos, still waiting to be tried.

69. An Improved Electrical Impedance Tomography (EIT) Algorithm for the Detection and Diagnosis of Early Stages of Breast Cancer (1998)[712]

Al Wexler (Figure 160) was both a Professor of the Department of Electrical & Computer Engineering at the University of Manitoba

Figure 160.   Al Wexler (1935–2021)[713]. (UC.)

and a businessman. He and I supervised Jerry Murugan for a PhD thesis in electrical impedance tomography which at that time seemed to have great prospects in imaging breast tumors[712]. Unfortunately, proprietary issues got in the way of continued collaboration. Once Natalie and I sat at his business to impress visiting Japanese businessmen with his number of employees, although we were not.

70. The axolotl as an animal model for the comparison of 3-D ultrasound with plain film radiography (1999)[714]

Gordon Kulbisky was then a resident in Radiology at the University of Manitoba. The conclusion says it all: "The axolotl's ability to regenerate entire limbs provides us with the opportunity to image bone and cartilage from their initial point of origin; thus, serving as an ideal model for future studies of cartilage and bone formation, such as that found in the developing human fetus"[714]. See also[715,716].

Chapter 5
2000s

71. The Emergence of Emergence: A Critique of "Design, Observation, Surprise!" (2000)[717]

There was a period during which I paid attention to artificial life[377,718,719], inspired by talks at the Digital Burgess conference[720] organized by Bruce Damer (Figure 234). Artificial life was an attempt to create computer creatures that seemed life-like. I wrote a critique of one of these papers[721], which was rejected by the journal in which it appeared. In effect, I argued that these authors "... take us back to the argument for God from design of Bishop Paley". I had read Darwin cover to cover at age 15 while bedridden with flu[722], and had a broad knowledge of evolution[723]. Not to be silenced, I published the critique in another journal.

One amusing sidelight of the Digital Burgess conference held in Banff, Canada is that I was the last one up the mountain to observe the Burgess Shale outcrop, and the last one down, accompanied by two patient participants.

72. Probe with Chest Shielding for Improved Breast MRI (2000)[724]

At the time the world needed a good breast coil for MRI breast imaging, in which I was peripherally involved. This goal having

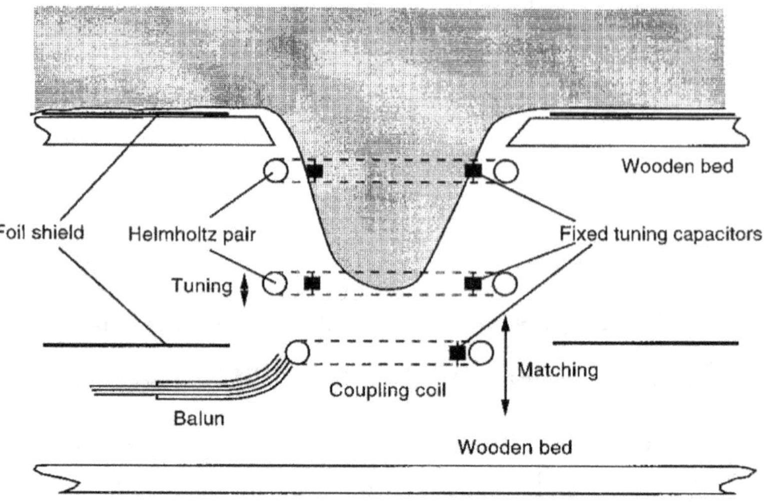

Figure 161.   Design of a breast MRI coil[724]. (With permission from John Wiley and Sons.)

been accomplished, I learned that my co-authors had no interest in seeing it tested clinically, much to my disappointment.

73. Adventures with a Business Administration Professor from 2000 on

I worked with Bryan Poulin (Figure 162) on a number of papers. These are so various that I don't select one as representative. We testified in the Canadian Parliament on the grant system, reviewed a book suggesting an alternative[649], and suggested how medical research in Canada might be organized[639]. We also criticized how science funding is done[637,638]. We discussed how to make the USA energy independent[650,725] (though we did not anticipate fracking problems) and discussed a world minimum wage[651,726].

Figure 162. Bryan Poulin. (With his permission.)

74. Books With Wings (2001)[727]

One day, shortly after 911 (September 11, 2001), I saw a television interview of the head librarian at Kabul University. At the end of the interview, he turned to the camera and pleaded for books. I said to myself, "I can do that". Thus began a 14 year episode of sending books to Afghanistan, mostly with the help of Canadian Forces[728,729]. It was over when they pulled out of Afghanistan[727]

Figure 163.   The logo for Books With Wings (2001–215)[727].

and a shipment from Tahir Shah of his father's (Idries Shah) books, from UK through Pakistan required a $10,000 bribe[730] to be released. At least 300 mostly medical students across Canada helped out, collecting, cataloguing and getting books to military depots, and the UK chapter sent books with the help of Spanish Forces[731]. US cooperation was minimal, despite efforts with Atifa Rawan[732]. All told, we shipped about $3 million worth of up-to-date books, probably now hidden or burned with the Taliban back in power. I am still friends with a few Afghans from that period. Sally Armstrong (Figure 164) wrote a chapter on Books With Wings in[733] and strongly supported the effort.

The books we sent were at most 10 years old at the time, so perhaps they're still useful, if hidden. Religious tracts sent by Iran kept the libraries warm in winters. I had requests from Ethiopians, but unfortunately, they were not at war with Canada at the time, so there was no way to arrange transport as we had with the books going to Afghanistan. I mailed them a few radiology books.

Figure 164. Sally Armstrong[734]. (Under the Creative Commons Attribution-ShareAlike License 4.0.)

A funny episode in Kabul occurred when Canadian troops made our first delivery, which they filmed. Afghan officials stood around, looking bored, as if they expected crap. (The USA had previously sent books on flowers in Arizona.) When handed a sample medical book, their eyes opened with joy.

75. Cytobots: Intracellular Robotic Micromanipulators (2002)[735]

Most micromanipulation inside cells is done optically. Here we designed robots small enough to operate inside cells. See also[736,737]. They have not yet been built and tested.

76. Electrorotation of Axolotl Embryos (2002)[709]

Natalie and I flew to the University of Calgary to work with Karan Kaler and his students Ghalia Abou-Ali and Reginald Paul on the effects of a rotating electric field on axolotl embryos. This phenomenon occurs when the axolotl generates electric fields at gastrulation. At neurulation, an electric field has been demonstrated in chick embryos[738].

Michael Levin is presently resurrecting bioelectricity[710]. But it has yet to be combined with our differentiation waves and molecular developmental biology.

77. A Simulated Comparison of Turnstile and Poisson Photons for X-ray Imaging (2002)[739]

A turnstile photon is one emitted by a quantum dot on command, as by a change in voltage. They are considered as components of optical computers. In this paper[739], with Cameron Melvin, I imagined that turnstile X-ray photons might someday become available.

Apparently, although some manipulation of the wavelength has been demonstrated[740,741], no one has yet to attempt to generate turnstile x-ray photons. As most X-rays have been generated using electrons impacting an anode, one might start with turnstile electrons[742–744] and accelerate them. The advantage is one of knowledge: if one knows that a photon is on its way, as opposed to standard x-ray sources, that produce x-ray photons at unknown times (with Poisson statistics), then one can improve the acquired image. It works. But not in reality until someone produces turnstile X-ray photons. Thus, this is a method for the future.

An alternative is to use quantum mechanically entangled photons, at least one of which is an x-ray photon, as suggested by Edward H. Sargent and described in[201]. The X-ray photon goes through the patient, while detection of the other indicates when. Again, this idea is for the future.

X-ray lasers with "multiplying" crystals have been invented. I never got into phase effects in X-ray imaging, but if an x-ray laser, which presently fills a large room, could be made smaller, it might find applications, especially in medicine. I wrote a letter to the US military when they dropped their x-ray ray-gun program, as it had medical potential. Got no reply.

In general, there are many other properties of X-rays that are not yet used for medical imaging, such as time-of-flight, polarity, and fluorescence (cf.[745]).

78. Reverse Engineering the Embryo: A Graduate Course in Developmental Biology for Engineering Students at the University of Manitoba, Canada (2003)[746]

I actually do like to teach, but preferentially with small classes. I've had undergraduate classes from 8 to 1500 students. In the latter,

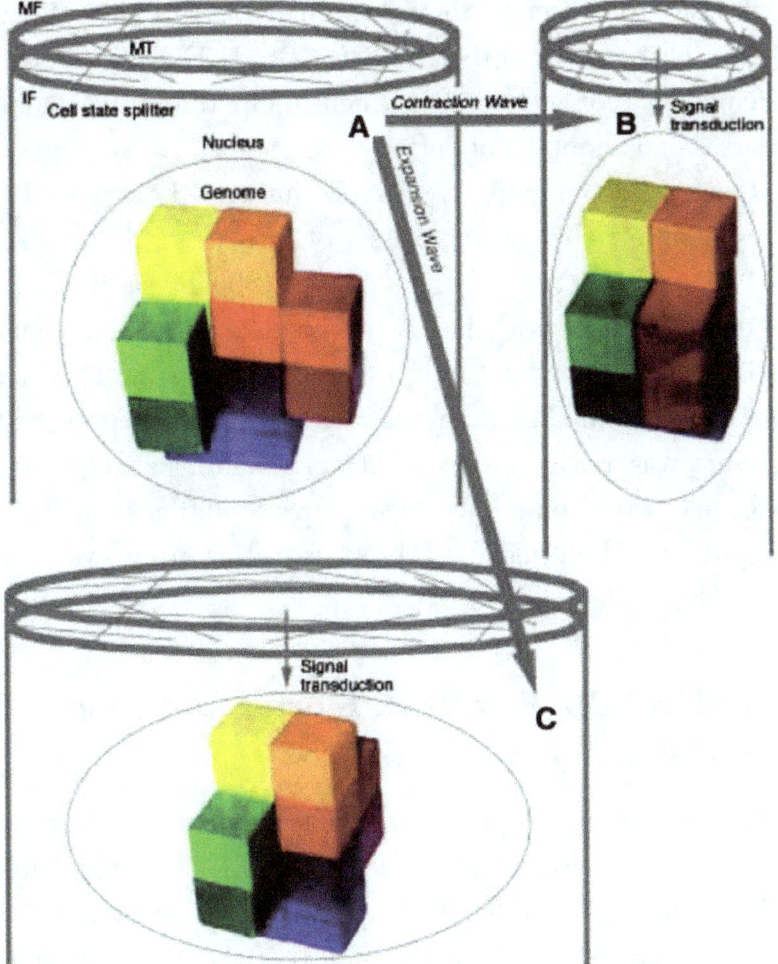

Figure 165. A toy model of the cell nucleus as a Wurfel[746]. (With permission of the *International Journal of Developmental Biology*.)

I once assigned a two page essay, and after reading 50 on epilepsy, imagined I had that condition. It took me a week full time reading to grade them. The class that I enjoyed the most was a class of 30 education majors who planned to teach elementary school. I had them do biographies of biologists and took them on field trips.

I came to regard understanding embryogenesis as reverse engineering of the embryo. This paper[746], with engineering student

Melvin Cameron, discusses how we went about this (cf.[747]). Melvin was a charter member of the IEEE EMBS (IEEE Engineering in Medicine and Biology Society), which I helped found. My graduate classes had about 4 students.

Teaching assignments were very uneven. I suggested that faculty members could teach classes of 30 instead of one teacher crowd controlling 1500 (of which only the 50 who sat in the front rows expressed interest), but it fell on deaf ears. Some faculty complained of having to teach for 2 hours per year. The faculty union, which I joined on their board as the medical representative for 6 years, was mostly interested in picketing for social issues, and during that period only once heard a presentation by a student; afterwards they belittled her. My ideas in this group were but a thorn in their sides.

79. 17th NADS, Key West, Florida, October 21–26, 2003

NADS (North American Diatom Society) has fun meetings every couple of years held in isolated locations, such as field stations. I've been to a few. Evelyn E. Gaiser and I organized this one, introducing diatom nanotechnology[748]. To make sure that the nanotechnologists and mostly diatom ecologists stayed to the end, we deliberately alternated their presentations, rather than having separate sessions. It worked, and they got to know one another. Later, we did publish a journal issue on diatom nanotechnology[749], including[704] one of the organizers who couldn't make it to Florida: Frithjof Sterrenburg (Figure 166), who was a superb amateur diatomist whom I later visited at his home in the Netherlands[284,750].

Figure 166.   Frithjof Sterrenburg (1934–2016). (Source unknown: not on Internet.)

80. Experiments with the Nonlinear and Chaotic Behaviour of the Multiplicative Algebraic Reconstruction Technique (MART) Algorithm for Computed Tomography (2004)[751]

It has always amazed me how rich the problem of computed tomography can be. In attempting to improve the MART algorithm, Cristian Badea (Figure 167) and I ran into a brick wall: the chaotic behavior when the correction term was taken up to the power 2, instead of >0 to 1.999[751]. Our conjecture of chaotic behavior was proven mathematically by Tetsuya Yoshinaga and colleagues and the improved results until the power of 2 was reached were further investigated[752–762]. As any MART algorithm automatically confines the image reconstruction to the convex hull, rather than throwing data to the infinite plane, as in Fourier algorithms, Power MART may become the algorithm of choice. We did not explore power >2, which might produce surprises.

Figure 167.   Cristian Badea. (With his permission.)

Figure 168.   Anne Ussing. (UC.)

Sitting in on an engineering course on chaos by Witold Kinsner prepared me to recognize this behavior, and an earlier time when I learned about chaos in heart problems.

I also found chaos in the movement of the colonial diatom *Bacillaria*[263,763]. Anne Ussing (Figure 168), a science librarian who I visited in rural Denmark, did the translation of an archaic Danish

1700s paper on *Bacillaria*, which was thought then to be a "peculiar" beach animal. Anne also wrote books on horse pigmentation and published research on axolotls and mice.

81. 3D Breast Image Registration — A Review (2005)[764]

I was interested in the problem of image registration for a while with graduate student Radhika Sivaramakrishna (Figure 169) coming up with a novel "Starbyte" image transformation[765–768], Xiohua Zhou (Figure 60) used geometric unwarping[769], and Andrzej Mazur used simulated annealing[770], but I stopped when it became a crowded field and it became clear that it is better to hold the breast stationary between diagnosis and cure if possible. We ran two workshops that we called "Alternatives to Mammography", hoping they would run every year until breast cancer was conquered[771],

Figure 169. Radhika Sivaramakrishna. (With her permission.)

but interest petered out. Other work on breast cancer imaging includes[256,772–774] and comments[775–778], including a demonstration that *Mammograms are Waldograms*[779], distributed widely at a Radiological Society of North America conference in Chicago with 90,000 participants.

Her husband Shashi (Figure 165) came with her from India as a postdoc[780,781]. They were used to servants in India, so we had to teach them how to boil an egg and clean a toilet. Productive and kind, they still had to learn about our ways. Culture shock goes two ways.

Two amusing sidelights are that I had some medical students get cow udders from a local butcher and X-ray them. They could see no internal structure, because these were young cows with dense udders. It's hard to get an old cow.

In another *faux pas*, I made a "phantom" from various plastic pieces suspended in clear silicone. A student almost destroyed a $12,000 x-ray source trying to get an image. I had forgotten that visually clear silicone is x-ray opaque. Close call.

Figure 170.   N. S. (Shashi) Shashidhara. (With permission of Radhika Sivaramakrishna.)

82. Local Independence in Computed Tomography as a Basis for Parallel Computing (2005)[350]

One of the remarkable features of CT is that local manipulation of a CT image hardly affects parts of the image at a distance, even though the x-rays can crisscross in many directions[344]. Dan Martin investigated this property with Parimala Thulasiraman, a Professor in Computer Science, University of Manitoba. I have long been interested in the prospects for parallel computing in CT[782,783] since the 64 processor ILLIAC IV[784] and experimented with distributed computers. Much more parallel processing capacity is now available in some desktop computers.

The relationship between local manipulation of a CT and local tomography[555–557] has not yet been explored.

83. Can We Stop AIDS in Afghanistan? (2005)[786]

I met Wassay Niazi (Figure 172) when he joined Books With Wings[727] in Winnipeg. He had been an Associate Professor of

Figure 171. Parimala Thulasiraman[785]. (With her permission.)

Figure 172.　Wassay Niazi with two of his young daughters who escaped Afghanistan. (With his permission.)

Infectious Diseases at Kabul Medical Institute, and as a refugee was unemployable in Manitoba medicine despite his experience with tuberculosis, much needed then in Canada. He and I tried to thwart the HIV/AIDS epidemic in Afghanistan, when there were only a few cases, hoping to nip it in the bud. We tried to bring in experienced people from Uganda to no avail, even though we had members of the Afghan government involved. Our grant application to the World Bank was turned down. Our impression was that no one listens when they would rather repeat their failures elsewhere (as in Pakistan) than give upstarts an opportunity to try. We gave many talks, such as[786]. Wassay moved from Manitoba to Alberta, where he could find a job that used his medical skills.

At the time of the abandonment of Afghanistan by the USA, it took 11 of us to get Jawad Kia (Figure 173), Books With Wings

Figure 173.   Jawad Kia and his daughter. (With his permission.)

volunteer in Afghanistan, and his immediate family on the last rescue plane out of Kabul: 2 in India, 5 in Winnipeg, and armed escorts in Kabul. Our main obstacle was helping him fill out the federal Canadian paperwork which used antiquated software, while he had only occasional electric power.

84. Grasping for Wholeness: A Review of Stuart Pivar's book *Lifecode: The Theory of Biological Self Organization* (2006)[787]

It is rare that an author one reviews bites back, but this one is the closest I came to a lawsuit[788]. Natalie and I were hired and received a hefty fee of which half was paid up front, to review his book. To our astonishment, when we submitted the review to him, he threatened to sue us even though the review had not yet been published. I sought the intervention of Judy Anderson, then an Assistant Dean of Medicine, who deftly avoided the problem. We never got paid the other half of our fee. We cannot remember why it lacks Natalie's co-authorship. At present, in embryogenesis and origin of life, one could still say that we are grasping for wholeness.

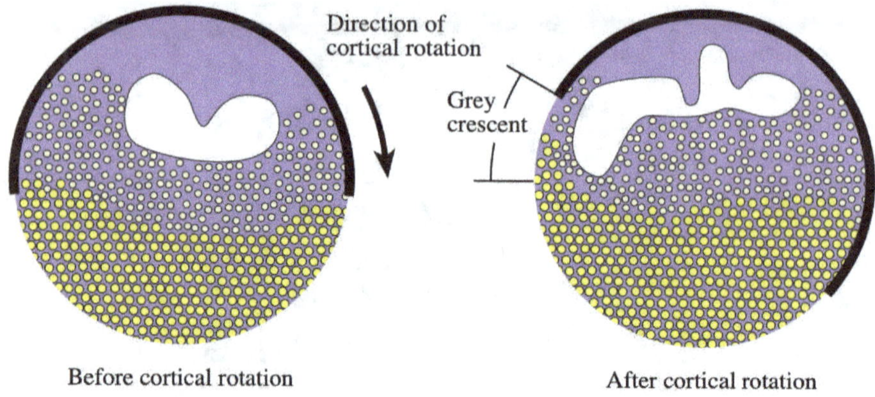

Figure 174.   Cortical rotation of the fertilized axolotl egg[789].

85. Yolk Dynamics in Amphibian Embryos (2006)[789]

I gave lectures to a group of mathematicians[789–792] who decided to tackle the motion of yolk during embryo cortical rotation. Afterwards some of them went off on the problem on their own[793]. Previous work:[794].

86. A Hypothesis Linking Low Folate Intake to Neural Tube Defects due to Failure of Post-translation Methylations of the Cytoskeleton (2006)[795]

This hypothesis by Natalie Björklund-Gordon about these human birth defects and their prevention has held up over the years and may lead to a reduction in neural tube defects in humans, as in the following reviews:[796–802]. This work was completed during her PhD thesis in the Department of Human Genetics at University of Manitoba, (now Biochemistry and Medical Genetics.) Her insistence on completing her PhD slowed down but did not stop our collaborative work.

87. The Number of 3D Configurations of a Labeled Size 2*n "Wurfel" (2006)[803].

The Wurfel was a toy model that we used to show how the nucleus of a cell in an embryo could have many 3D configurations (Figure 175), each possibly representing a differentiated state of the cell[149,183]. I therefore was curious if this toy actually filled the bill, and approached mathematician John Tromp (Figure 176), who did the calculation[803].

n	0	1	2	3	4	5	6	7	8	9	10
#	0	0	1	8	22	256	2,247	21,576	225,102	2,303,014	24,563,283

11	12	13	14	15
267,169,300	2,937,239,494	32,814,269,626	370,231,763,542	4,215,902,111,928

Figure 175. The Wurfel in three configurations. The top configuration was not considered in Tromp's calculations, so the table gives minima.

Figure 176.　John Tromp[804]. (With his permission.)

It can be seen that the number of configurations increases rapidly with the size of the Wurfel, far surpassing the number of cell types in an embryo as it develops. While the Wurfel is just a toddler's toy, it shows how it might be possible for all cells in a developing embryo to contain the same DNA, but be of many differentiated types.

A similar lesson may apply to the nuclei of atoms (Chapter 11 in[73]).

88. An Iterative Three-dimensional Electron Density Imaging Algorithm using Uncollimated Compton Scattered X-rays from a Polyenergetic Primary Pencil Beam (2007)[805]

This was the graduate thesis work of Eric Van Uytven (Figure 177), who I helped supervise. See also[806–808]. Eric is now a Medical Physicist.

Figure 177. Eric Van Uytven. (With permission of CancerCare Manitoba Communications.)

89. Diatoms in Space: Testing Prospects for Reliable Diatom Nanotechnology in Microgravity (2007)[809]

Since diatom nanotechnology was taking off[284,358,812–814], applicable from drug delivery to solar panels, I surmised that if one cell of

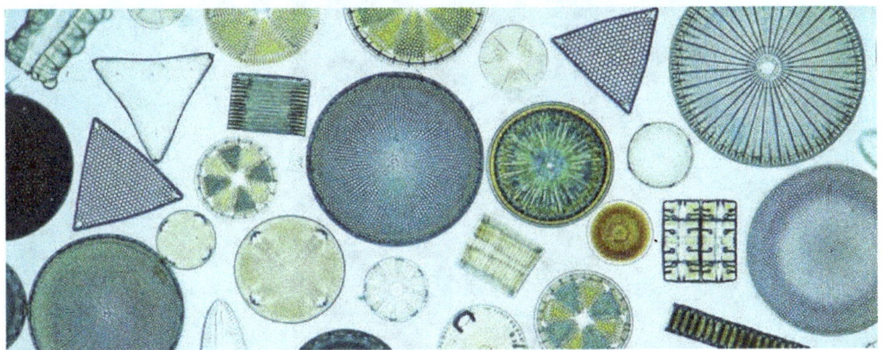

Figure 178. Some of the variety of size and shape available across diatom species. (From[810] with permission of Rob Kimmich.) See also[811].

each species (maybe 200,000) were taken aboard a spacecraft, a tiny package, the whole range of diatom nanotechnology would be available to astronauts.

Our interest in microgravity extended to embryos, with a pediatric lecture[815] that included the motto: "To boldly breed where no one has bred before!". My graduate student Susan Crawford-Young (Figure 186) reviewed microgravity in biology[816].

90. Could There Have Been a Single Origin of Life in a Big Bang Universe? (2007)[817]

Using a guesstimate of when the universe cooled down enough for some parts to sustain life, the radius of the universe at that time, and my estimate then of the maximum speed of life carried by astronomical ejecta, I defined a "biohorizon" by analogy with the light speed horizon for the universe. The result was that, if we presume that the universe is full of life, it would require 50,000 origins at that early time to accomplish that[817]. In light of major revisions of my speed estimates, this calculation needs updating. Richard Hoover, my coauthor (Figure 179), is a retired NASA

Figure 179.   Richard Hoover. (With his permission.)

astrobiologist and diatomist, who has published much on possible microfossils in meteorites, and life in comets. I have critiqued the "evidence" that stratospheric diatoms come from space[818,819], and the porosity of meteorites allowing earthly microbes to crawl in[69], which I think we can count as friendly sparring. We have cooperated on two papers[809,817]. He was the first person to attempt to put diatoms in a spaceship.

91. There is but one Journal: The Scientific Literature (2008)[348]

While some scientists essentially publish the same papers in as many places as possible, I have taken the attitude that with current search tools once is generally enough. This attitude of mine makes for poor self-promotion.

92. Monotheism: The Basis for Unifying Abrahamic Religion and Science? (2008)[820]

Much has been written on the relationship between science and religion for over a century. I got into this when I first met Joseph Seckbach at a meeting in San Diego. He asked me for editorial help. I responded by asking for 6 months to get dialogue going via e-mail between the chapter authors. Most of this was printed verbatim at the end of each chapter[821]. Two were not polite, and were deleted, one on each side. My comments on monotheism, based on the history that much of science grew out of trying to figure out how God worked in the Universe, were both my talk to our synagogue and this preface. See also[822].

The book included creationists and anti-creationists[823]. Sometimes I made fun of both[824]. For instance, making up my own rules, I could show that a sequence of letters "so-and-so is wrong" occurred 5 times per page in an anti-creationist's chapter. As a kid, I was brought to a synagogue once, the Bahai temple in Chicago once, and once went to a Catholic church with a girlfriend. I was raised as a tolerant atheist, and remain so. To me, all belief is tentative, based on our current best interpretation of evidence, though that is often flawed by preconceptions.

Other books[825-827] and chapters[828,829] followed with Joseph. Joseph produced about 30 books for Springer, who terminated him as an editor when two French authors complained, *after* production of a volume edited by both of us, that they couldn't be seen in the same book as religious authors. We tried Elsevier for a while[60], had some troubles, and ended up with Martin Scrivener (Figure 249) of Wiley-Scrivener, with whom we've had productive series of books on diatoms and astrobiology.

At any rate, our perception of this vast universe or multiverse changes almost daily, with no evidence so far of life elsewhere nor of Heaven. As a scientist, I suspend judgement.

93. Rayleigh Instability of the Inverted One-cell Amphibian Embryo (2008)[830]

The result of inverting the fertilized axolotl egg is either normal development or none. I figured this was due to the dripping either coming down the side or middle, respectively[568]. Working with experts at simulating fluid flow, Roel Luppes (Figure 181) and Arthur Veldman[831], Roel, who flew from The Netherlands, and my postdoc Comron Nouri (Figure 182) examined the Bessell functions for this situation. These are either asymmetric, corresponding to dripping down the side, or symmetric, dripping down the middle.

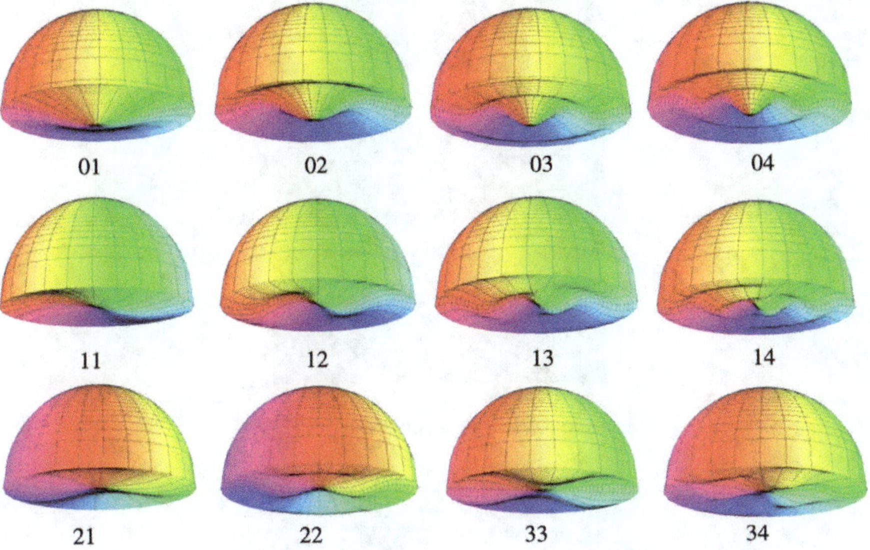

Figure 180. When a 2 mm salamander egg is turned upside down, the heavier yolky cytoplasm is placed on top and drips down[830]. (With permission of IOP Publishing.)

Figure 181. Roel Luppes[832]. (With his permission.)

Normal development thus involved a rotation of the cytoplasm relative to the membrane of the egg, to which we attributed alignment of the cortical microtubules[833,834]. Comron Nouri (Figure 177) was a postdoc who worked with me at the University

Figure 182.   Comron Nouri[838]. (UC.)

Figure 183.   Jack Tuszynski. (With his permission.)

of Manitoba and then Jack Tuszynski at University of Alberta (Figure 183). With Jack, who has delved into a possible microtubule basis for consciousness[835], I suggested that consciousness of the one-cell axolotl embryo be explored[834,836]:

"Finally, we wish to indulge in some speculation regarding the highly debated relationship between microtubules and consciousness[835]: 'One hemisphere of the cortex of the one cell axolotl embryo maps through development to the brain. Therefore there may be some continuity between the microtubules of the one-cell embryo cortex and those of the brain. If quantum coherence occurs in the brain, why not in the one-cell embryo?"{Tuszynski, 2008 #106508.

The one-cell salamander embryo is much more accessible than one's brain. Maybe someone will explore its quantum mechanics and its degree of consciousness someday[837].

94. Potential of Silica Bodies (Phytoliths) for Nanotechnology (2009)[839]

Phytoliths are diatom sized specks of silica that are found in most plants, and are believed to be responsible for grass' ability to grow tall. The stems of most crops are just burned or plowed under, throwing away a potential vast source of natural nanotechnology. On mentioning this prospect to a Manitoba Government agriculturist, his eyes just glazed over. Manitoba is a major wheat and thus chaff producer and burner. Suresh Neethirajan was a graduate student in Biosystems Engineering at the time, one of the departments I was in for a while.

Figure 184. Suresh Neethirajan. (With his permission.)

This paper has been cited by a number of papers on phytolith nanotechnology, mechanics and optics, and seems to have awaken some people to the wasted potential.

95. Google Embryo for Building Quantitative Understanding of an Embryo as It Builds Itself: I. Lessons from Ganymede and Google Earth (2009)[840]

Before Google became the world leader in censorship, such as on YouTube and its search engine, it aspired to make the whole world available via Google Scholar, Google Books, Google Translate, Google Images, and 3D maps such as Google Mars. As I was trying to image whole, nearly spherical, early axolotl embryos[841,842], I conceived of Google Embryo[840,843]. Two fellows from Google showed up

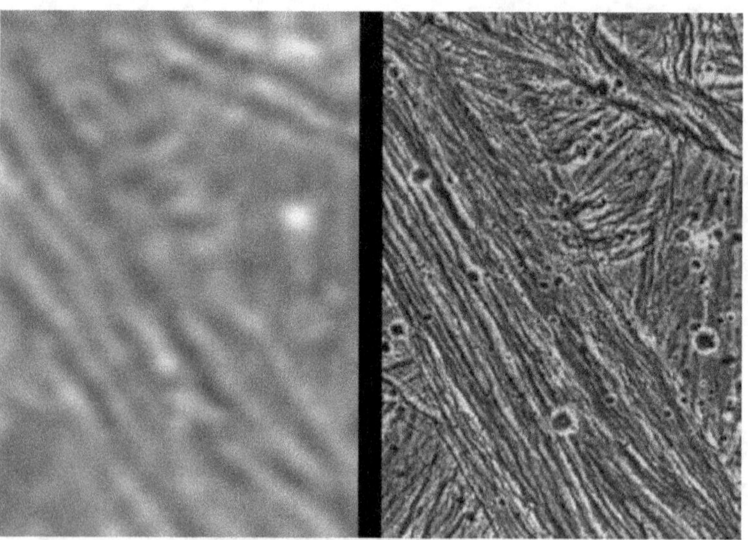

Figure 185.   Improvement in imaging of the surface of Ganymede, 1979 to 1996[840]. (With permission of Springer Nature.)

Figure 186.   Susan Crawford-Young. (With her permission.)

while I was a visiting professor at MIT, but then dropped out saying Google stopped allowing its employees to collaborate. This attitude was quite contrary to the business success stories that Bryan Poulin (Figure 162) and I had documented[639]. In retrospect, it was the beginning of the decline and maybe coming fall of Google. With the Hunter Biden affair, I was later personally censored by Google, and immediately cancelled my paying Gmail account, losing more e-mails than Hilary. I now only use their old services, and avoid any others that require an account with them. Susan Crawford-Young (Figure 186), my former Master's student[844], carries on with 3D/4D imaging of embryos[494,816], introducing optical coherence tomography[845].

96. Milking Diatoms for Sustainable Energy: Biochemical Engineering versus Gasoline-secreting Diatom Solar Panels (2009)[846]

I had a trip to Moscow, Russia to attend an embryology conference, when I received an invitation from Unilever Hindustani to give a talk on diatoms at their company's annual meeting in Bangalore,

Figure 187.   The diatom with an oil droplet that started it all[846,847] by Charles O'Kelly. (UC.)

India[281]. They paid to reroute my airplane rides, and put me up, in what turned out to be a palatial room with all granite furniture. I asked a graduate student there, Karthick Balasubramanian, to bring some microscopes and live diatoms, so my audience could see some moving diatoms. After my talk, he arranged for me to visit his professor, T.V. Ramachandra (Figure 188), who was in the Energy & Wetlands Research Group, Centre for Ecological Sciences/Centre for Sustainable Technologies, Indian Institute of Science. I had one slide showing a drop of oil inside a single, live diatom (Figure 182). When he saw this, his eyes lit up, and this paper with him and Karthick became my conference report[846]. It was clear that somehow, diatoms could replace fossil fuels[848], much to the chagrin of the greenies, and we could keep our gasoline

Figure 188.   T.V. Ramachandra. (UC.)

engines, perfected for over a century, with net zero carbon foot-print. (Greenies don't want actual solutions to their shibboleth of Global Warming, now called Climate Change.)

I got into the idea, then looking at the runways for algae (easily contaminated; I visited a runway company, now defunct) and photobioreactors (expensive) even presently used for growing algae *en mass*, came across the small literature, started by a high school kid for a science fair, of growing microorganisms inside bubble wrap. When manufactured, apparently the insides of the plastic bubbles become/remain sterilized. I brought the idea to Sealed Air, the company that started bubble wrap. My host's manager refused to speak with me after my talk. (Cost me US$1000 to get there.)

I wrote a paper on what I called Bubble Farming[849]. Vandana Vinayak has since carried the ball[850–854], finding a diatom that spontaneously secretes its oil[855,856], reviewing the prospects for diatom oil[854], and further designing diatom solar panels[857]. Cliff Mertz[858] has also tried this. With some jars at home, and pond water, I demonstrated that the plastic used in air pillows retained water for months, while allowing oxygen and carbon dioxide across the plastic. An optimal kind of plastic with these properties has yet to be sought, and Bubble Farming needs to get out of the lab into agricultural testing. Optimal choice of a diatom species or other microalgae or cyanobacteria for Bubble Farming, perhaps with their genetic engineering, has yet to be researched.

97. *Hyalodiscopsis plana*, a Sublittoral Centric Marine Diatom, and Its Potential for Nanotechnology as a Natural Zipper-like Nanoclasp (2010)[859]

The highest-level structure achieved by diatoms is the colony[860] (unless we count diatom mats). In this colonial diatom the diatoms

Figure 189. A diatom zipper[859]. (Open access.)

Figure 190.   Ille Gebeshuber, (From[861], in Public domain.)

hold together like a zipper, flexible but strong. How do they build this? Ille Gebeshuber[861] (Figure 190) is an engineer interested in tribology[862] (friction) and other aspects of diatoms[268,863–867].

98. CS: ART and CS in CT, Regularization for Matrix Completion, Estimation of High-dimensional Low-Rank Matrices (2010)[868]

There are two developments I tried to stretch my brain around, but couldn't quite make it: compressive sensing (CS)[869–873] (Figure 199) and machine learning (ML)[874]. So, I just got involved with others who knew these areas to see what happened. Whether these approaches are passing fads or centrally important remains to be seen.

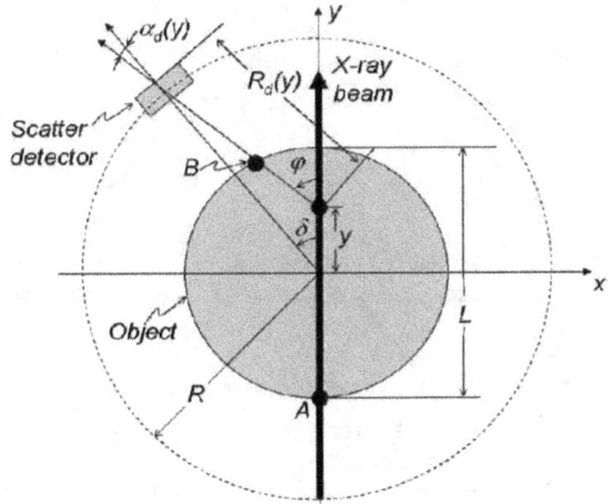

Figure 191.   A 4π detector for scattered x-rays[773]. (With permission of IOP Press.)

99. A Novel Hybrid Reconstruction Algorithm for First Generation Incoherent Scatter CT (ISCT) of Large Objects with Potential Medical Imaging Applications (2011)[773]

A configuration was proposed that would capture all scattered x-rays in all directions (4π steradians). See also the next section. A 4π walk in configuration was proposed for PET (Positron Emission Tomography), shown to me by Michael Neumann at the University of Chicago, brother of John von Neumann[875], which might have inspired it.

100. A 1st Generation Scatter CT Algorithm for Electron Density Breast Imaging which accounts for Bound Incoherent, Coherent and Multiple Scatter: A Monte Carlo Study (2011)[774]

This paper is further work on 4π imaging. In ordinary breast CT, substantial and multiple scattering of x-rays occurs in the breast.

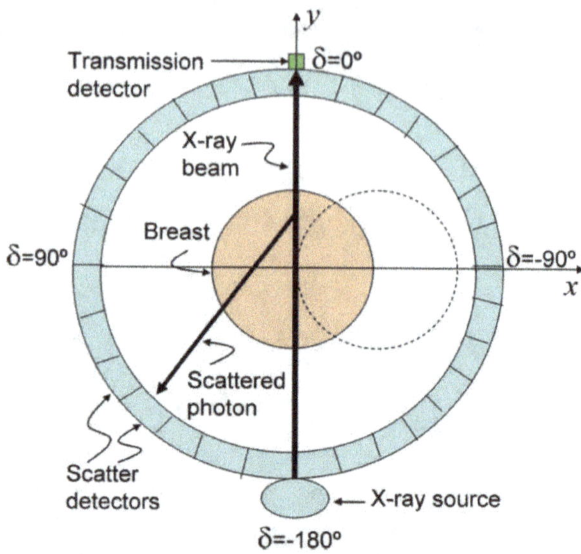

Figure 192.   Configuration for scatter detection from a pendent breast[774]. (With permission of IOS Press.)

Figure 193.   Jorge Alpuche Avilés[876]. (With permission of CancerCare Manitoba Communications.)

Jorge Alpuche Avilés (Figure 193), then a graduate student from Mexico, led this attempt. He is now a Medical Physicist.

101. Epilogue: The Diseased Breast Lobe in the Context of X-chromosome Inactivation and Differentiation Waves (2011)[877]

The possibly most important aspect of this paper is the testable prediction of where and when stem cells arise in embryos. Lu Kai (Figure 195) and I have further discussed stem cells in the context of differentiation waves[878–880].

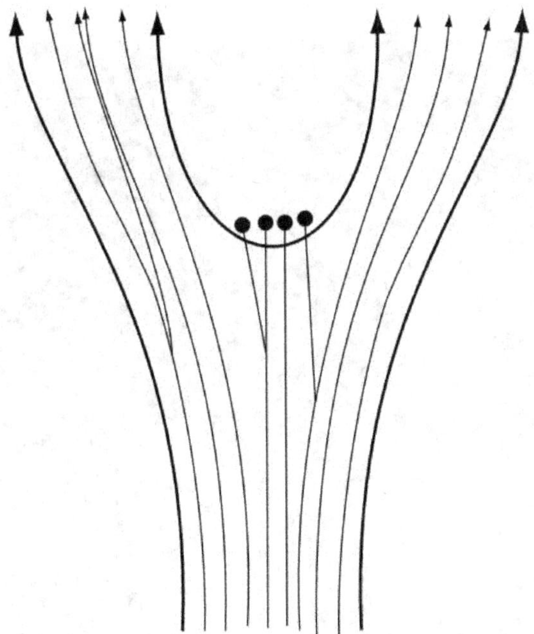

Figure 194.   The heavy lines indicate two pathways that cells can take at a branch of a differentiation tree. The lighter lines refer to the cell lineage. Their branch points are individual cell divisions. However, this paper hypothesized a third set of cells that took neither branch of the differentiation tree, but paused further differentiation as stem cells. If correct, stem cells would arise at the boundaries between differentiation waves. (With permission of Springer Nature.)

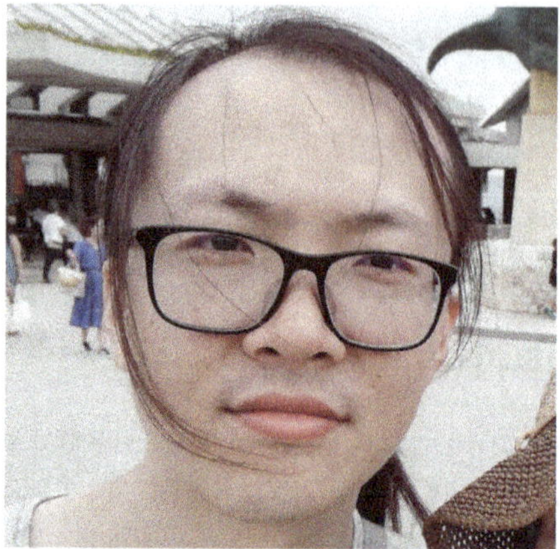

Figure 195.   Lu Kai. (With his permission.)

Figure 196.   Cosmic embryo[882]. (Public domain.)

102. Cosmic Embryo #1: My Erdös Number Is 2i (2011)[881]

If one published with mathematician Paul Erdös, one had an Erdös Number of 1[883]. If you published with someone who published with Erdös, your Erdös Number was 2, etc. I was invited by mathematician Stan Ulam[884] (Figure 197), "father of the hydrogen bomb" with Edward Teller[885] to postdoc with him based on[79]. But we never published together. So, I took the imaginary Erdös number 2i[881]. While Ulam is touted as a great mathematical biologist,

Figure 197. Stanisław Ulam (1909–1984) "Unless otherwise indicated, this information has been authored by an employee or employees of the Los Alamos National Security, LLC (LANS), operator of the Los Alamos National Laboratory under Contract No. DE-AC52-06NA25396 with the U.S. Department of Energy. The U.S. Government has rights to use, reproduce, and distribute this information. The public may copy and use this information without charge, provided that this Notice and any statement of authorship are reproduced on all copies. Neither the Government nor LANS makes any warranty, express or implied, or assumes any liability or responsibility for the use of this information."

I was bent on solving biological problems. It became clear to me that Ulam only looked to biology as a source of interesting mathematical problems. He regarded my work at that time on polysomes[186] as mathematically trivial, and thus the impasse. Perhaps these discrepancies in motivation are widespread?

Ted Puck (Figure 36), head of Biophysics at the University of Colorado, hired me to be the go between, between him and his friend Ulam, who took me on unseen based on my "snail" article[79]. While I learned a lot in that year of my first postdoc, especially about chromosomes[877], as a go-between I was a failure. It became clear that Puck knew no mathematics, while Ulam was not versed in biology, so the assignment was impossible.

I met Puck some years later, in his 90's, and only recall our mutual chagrin that Jack (John R.) Sadler had died young in a rock climbing accident[886,887]. I knew Jack back at the Institute for Molecular Biology at the University of Oregon, and sort of followed him to Colorado. Unfortunately, we never found something in common to work on. I recall going rock climbing in the Rocky Mountains once with him and a bunch of boys. They had to haul me up on a rope to the pinnacle.

During my year at the University of Colorado, I attended a course at NCAR (National Center for Atmospheric Research in Boulder) on laser profiling of clouds, which in retrospect I can see was an aimed beam method somewhat analogous to what we're now trying for breast imaging, except that it depended on the timing of the returned light to profile clouds. Maybe someday we could do such time-of-flight measurements for X-ray photons.

My blog was named "Cosmic Embryo" after an embryo shaped galaxy[882] (Figure 194).

103. Cosmic Embryo #3: The ART of 3D Sun and Breast Cancer Imaging (2011)[888]

I knew that the ART algorithm was widely used in many fields. I pleasingly learned that two satellites circling the Sun collected images from which the Sun's corona could be reconstructed by ART, until they were on opposite sides, and communicated by e-mail with a scientist involved. Now we get daily videos of the Sun and its flares. I did later meet Joshua Lederburg[889], as I was interested in his Dendral system for computer representation of organic molecules (which I amateurishly surveyed in high school), and much later suggested that taxonomy of multicellular organisms be based on differentiation trees rather than ill-defined characteristics[183]. If I'm right, it has become obvious that only later generations might rediscover this possibility[183].

104. Bioinspiration for Tribological Systems on the Micro- and Nanoscale: Dynamic, Mechanic, Surface and Structure-related Functions (2011)[863]

Tribology is "the study of friction, wear, lubrication, and the design of bearings; the science of interacting surfaces in relative motion", which harkens back to my obsession with *Bacillaria*. Thus, I tagged along with Ille Gebeshuber (Figure 190), an Austrian engineer expert in diatom tribology. I reviewed diatom nanotechnology from the point of view of patents[890]. I was invited to an art museum in Barcelona, Spain that emphasized Antarctic diatoms and art derived from them[864]. Janice Pappas has done an excellent job in recruiting artists for a forthcoming book on diatom inspired art, that I suggested[891], and taken over from me as a Series Editor for diatom books at Wiley-Scrivener.

At a 2003 meeting of the North American Diatom Society in the Florida Keys, Sarah Spaulding and I deliberately alternated nanotechnology talks with regular (mostly taxonomic) diatom talks so no one was tempted to leave early. This worked, and resulted in a special issue[704,749].

105. Triggered, Nanostructured Biodegradables (TNBs) for Surgical Implants (2011)[866]

On the way to a visiting professorship to MIT I got very sick. I had a gall bladder operation, perhaps from eating too much cheese-cake at an international diatom meeting *en route*. In Kingston, NY, as I only had medical travel insurance from Canada, I was told I would not be admitted to one hospital and would be forcefully ejected. A black janitor overheard this and kindly directed me to a public hospital. I thus had surgery there. My care was superb, especially when they discovered my well-known lawyer cousin lived nearby. He had sued several of the doctors and the hospital. They were wonderful about my care. I recall awakening with a very black surgeon standing over me, saying he had had his hands inside me. I just smiled. Shortly after we arrived in Boston com-plications arose from the surgery. I developed a constriction of the bile duct meaning another hospitalization and the installation of a special experimental stent. While this was regarded as an emer-gency procedure and covered by the medical travel insurance, its removal was not. We had quite a trial finding a way for removal of the stent while getting it covered by my Canadian health care. A Winnipeg doctor offered me 5-year survival if he couldn't get the stent out. Natalie arranged for a trip to Montreal. While visiting me she got lost in the Boston traffic, and calmed down after coffee and directions offered by a black man who noticed her plight. Having little else to do while in hospital, I wrote this paper about all the ways I could think about to have a stent disintegrate harmlessly on command so people wouldn't be faced with such issues. The gastroenterologist, David N. Schwartz, who referred me, was flattered when I acknowledged him in the paper. I was the healthiest patient on the Harvard Beth Israel Deaconess Hospital transplant ward and bought their tee shirt. My visiting professorship at MIT was not as productive as I would have liked it to be because I spent most my time struggling to function with a smoldering infection in the stent and constant pain whenever a movement pulled on it. We eventually drove to Montreal to have the stent removed. The surgeon in Montreal showed me how the stent had been flowing properly but was full of pus. However, my bile duct was saved. The constriction was gone. I immediately felt so much better that we walked back to our hotel and on the way we stopped at a deli. My appetite, gone for six weeks, was back. I had a delicious, hot Montreal smoked meat sandwich.

The reason for our trips south requires some explanation. Natalie had been having severe attacks with her lifelong asthma and in the winter of 2009/2010 as she finished a postdoc, she ended up in hospital six times. These winter difficulties with her breathing were an ongoing issue throughout our marriage but that winter was by far the worst. The last time was so bad that she had set up beside her hospital bed ready for when doctors and a respiratory therapist had to put her on a ventilator. Fortunately, she recovered enough that she was released later instead. In the past, she had always improved in the summer and spring was right around the corner. One of the doctors advised us we had to do something or she might not make it through the next winter. We bought a Ford F150 truck and a 30-foot travel trailer. We left our now college age children in charge of the house. I took a sabbatical beginning in 2010 September and we spent that winter in Arizona. I tied one of our dogs to a water pipe at a superb Mohave trailer camp, which he broke producing an Arizona geyser. Not only did this mean she did not have her usual horrible winter's trouble with her asthma, but she actually improved enough to be able to wean herself off most of the medications she had been taking for years. And so, part of my arranging the visiting professorship to MIT was to be able to take her south for a second winter. So, we left from Boston and headed south to what eventually became our first winter at Gulf Specimen Marine Lab, after having been intro-duced to Jack and Ann Rudloe, with whom we hit it off. We lived full time traveling around North America in that travel trailer for five years. Eventually we purchased our present home in rural Alonsa, Manitoba and continued traveling south for three more winters. Fortunately, after my own ill health meant no more travel-ing south she has made it through the winters in Alonsa with little to no trouble with her asthma. It was during these traveling years we finally finished our book, *Embryogenesis Explained*[149].

106. Octopus Art via Compressive Sensing (2012)[892]

For many years compressive sensing was of substantial interest, and I contributed many small items to the blog of Igor Carron (Figure 199), who was at the center of compressive sensing[893]. Due to need for home care of a child, he confined his work to this medium. Although we have never published together, Jack Rudloe (Figures 198, 200), author of many marine naturalist books[894], has

Figure 198. Jack Rudloe, 1971. (With his permission.). Photo by Bruce Montgomery, and his son Cypress Rudloe, by the author. (With Cypress' permission.)

Figure 199. Igor Carron. (With his permission.)

had a profound effect on my perspectives of the living world. We have been involved with oyster harvesting, shrimp in space, Water Worlds, the career of Ernest Just[895], a visit to his Gulf Specimen Marine Lab & Aquarium (GSML) by invertebrate embryologist Luigia Santella[896] (Figure 201), tall trees and hurricanes, artists, burying a sea water intake with gravel, dredging from a boat for specimens, sea turtle rescue, collecting huge quantities of the bryozoan, *Bugula neritina*[897] (for which I made up a ditty *Bugu*

Figure 200.   Jack Rudloe, President of Gulf Specimen Marine Lab & Aquarium, Panacea, Florida, collecting at low tide. (Photo by me, in Public domain [903].)

Figure 201.   Luigia Santella. (With her permission.)

la companie, a parody on *Vive la compagnie!*) and octopus intelligence. I filmed an octopus handing a toy giraffe back to Natalie, and one that seemed disgusted that her long hair draping into their aquarium had no tentacles. I pointed out to one young lady volunteering for Jack that except for accidents (hooks and propeller encounters), most turtles in north Florida needing rescue were cold stunned, contrary to the global heating mythology[898] and her proposed grant application. Once I violated Jack's rules and touched the dorsal fin of a nurse shark swimming in a circular tank, well away from its mouth. The shark responded instantaneously by splashing and soaking me. For 8 years Natalie and I escaped Canadian winters, until passing 75 my travel health insurance became prohibitive in cost. As a volunteer[899], I still perform occasional tasks remotely that Jack requests.

I edited a book, with a print run of 1, *The Living Jack: A Celebration of Jack Rudloe at 70 and His Living Dock*[900], with about a dozen personal stories of local friends of his that I knew at that time. While they were no literary giants, he treasured their gestures. I cribbed the title in part from his book[901,902].

One night, after Jack and I attended a boring meeting on how to attract tourists, I took a walk on that "Living Dock", and as I was admiring the brilliant stars above, "took a long walk off a shortened dock", falling at low tide into an oyster bed covered by a foot of water. Assessing that: 1) I was still alive; 2) hadn't drowned; 3) that the alligators that frequented the dock were elsewhere; 3) that I had many cuts and bruises from the oysters; 3) that it was a long walk in shallow water filled with Jack's experiments on growing oysters on trees, I had enough light to assess that my closest way out was up some stairs that I knew well from more delightful times on the Living Dock, watching busloads of kids of all races in the supposedly racial South collecting animals off its sides or from its motorized lift net, with Jack sitting amongst them. I inched my way over the gap in the dock holding onto the rail for support, and slowly made my way back to Jack's beachside home (elevated for attempted hurricane proofing), dripping blood all over his floor. Since Natalie had had medical training, I suggested he fetch her from our nearby RV trailer, and I was properly cleaned up in his incarnadined tub, bandaged up, and given an antibiotic for ubiquitous marine infecting bacteria. What had happened was that the kid doing repairs on the dock placed no barriers in front of his incomplete repairs, not anticipating star gazers. I think this was commemorated for a while (perhaps until the next hurricane wrecked the dock), with a plaque "Dick leapt here". In an attempt to retrieve my glasses, Jack checked his flashlight by aiming it at his one good eye and blinded himself. The kid did the retrieval. I considered suing Jack and then giving him the money, but that made no sense. I guess that my thinking patterns escaped me for a while.

One comment on the "racial South", besides the mixed teachers, students, and Jack's volunteers, is an amusing observation when we needed our truck's oil changed. A white man brought his pickup into the next bay, and the black attendant remarked that the Confederacy flag flying in the truck was part of the man's culture, when the latter tried to apologize for the apparent affront. They then both noted they were voting for Trump. Contrary to the fake news of increasing racialism of the USA, as shown above even in the North black people treated us with respect and help where needed, which is why I counter by mentioning them explicitly.

I took a photo of Jack's son Cypress (Figure 198), who runs GSML, in the posture of Jack at his age. The resemblance is remarkable. Cypress loves his life, while his friends he grew up with are sitting in office jobs. He married a South Korean lady and is raising a son who might be interested in keeping GSML going another generation.

While I have never delved deeply into animal behavior, I have been amazed how smart, like that shark, they seem to be. A wild shark followed me as I walked along the Saint George Island beach looking for beached Portuguese-Man-of-War for Jack. I kept my distance and it didn't get a meal. I once confined an ant within a ring of kerosene. After determining that it was surrounded in all directions, it just plunged through the ring. Not bad for having a few hundred brain cells!

107. The Vanishing Physician Scientist: A Critical Review and Analysis (2012)[904]

The vanishing physician scientist was my last paper as a University of Manitoba Professor. It was an assignment I was given by my Department Head of Radiology to be completed during the winter in the year of my visiting professorship at MIT and then our trip to Florida. Though I was no longer on sabbatical I had no specific teaching or committee duties at the time. I made a point of attending meetings virtually as required. My department head was very understanding about my wife's health issues and he wanted this review completed and assigning this while I was away benefitted us both. I had numerous other projects on the go and produced several publications. He and one other member of the department I was working with were supposed to be coauthors on this one. I am not sure what they expected from me. I think I was supposed to be some kind of magical solution to their problem of not having many new medical students showing any interest in research. I concluded that the book I reviewed[905] was a plea for special funding of one category of scientists, physician scientists, and that most medical admissions committees screened out the very people who were most motivated to become physician scientists. This alarming conclusion flew squarely in the face of how proud the field of medicine in general was about moving to a kinder, gentler and more understanding type of physician. My coauthors insisted on removing their names and refused to have anything more to do with the paper. We had a change of Dean while I was away and the entire feel of the place had changed. The new dean was adamant it was time to remove a lot of the "deadwood" in the university, i.e, faculty who didn't bring in lots of grants. Everyone was afraid they would become victims of the great makeover.

I soon became part of the designated "dead wood". Shortly after my return to Canada I was sent a formal letter advising me I was henceforth "on call" by the department 24 hours a day, seven days a week, three hundred and sixty-five days a year. Being placed "on call" was a maneuver normally reserved for physicians during emergencies like the Red River Flood of 1997. However, it meant that at any time the department head could call me and I had to be able to present myself to him, in person, within twenty minutes or be fired. The dire potential emergency I was on call for was if a medical student required emergency assistance with a radiology research project. I consulted with my union. I considered legal action. I was told it would take a year or more during which time I would have to remain in Winnipeg, on call, or risk being fired and losing my pension and benefits. Natalie would not get her winters away from her asthma troubles. In the end, at age 67, I decided that this was not a battle worth fighting. I proposed giving them the required six month's notice and that I was retiring on the condition they removed my on-call status. The new Dean, triumphant that his probably illegal maneuver had freed up more of his salary line by driving out yet another dead wood professor, immediately agreed. I was grudgingly offered a proforma Emeritus Senior Scholar position with distasteful strings. Already having a bad taste in my mouth over the "on call" maneuver, I declined. There was some grotesque nonsense about confiscating the last of my grant money for cleaning nonexistent hazardous chemicals in my laboratory space. I got around that by having the funds transferred to another department to the colleague who was a co-applicant on the grant. It was a bitter end to what had been a wonderful and productive career as a member of the Radiology Department. There was no retirement party, no mention in departmental newsletters or anything else to mark my departure from the University of Manitoba. I was simply vanished, and no one was told. In the following years Natalie or I would occasionally run into an old colleague and be asked what happened. In the end it turned out to be a blessing that they forced me out when they did because my retirement package included lifetime access to the university library. This wonderful privilege allowed me to continue my research. That privilege was rescinded 6 months later for all subsequent retirees. I didn't take it personally because I was not the only "deadwood" so removed by the new dean and I already knew institutions like universities have no memory.

108. A Mean Field Ising Model for Cortical Rotation in Amphibian One-cell Stage Embryos (2012)[834]

Ising models have figured prominently all my life. In this paper, we used it to model the first step in formation of the brain. Jack Tuszynski had a serious look at the possibility that microtubules are the basis for consciousness[835].

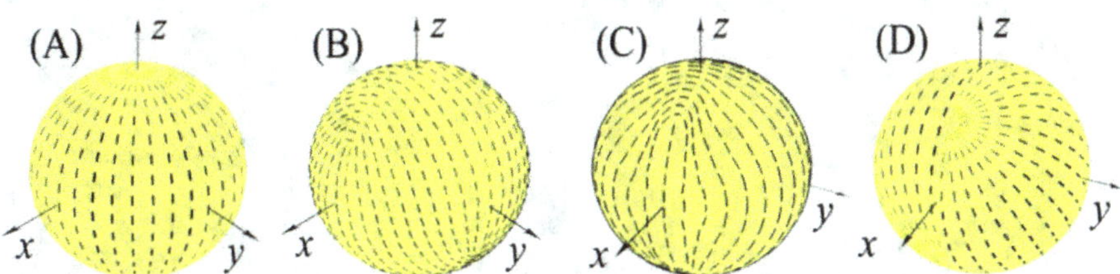

Figure 202. Some possible arrangement of microtubules on a spherical embryo[834]. (From[906] with permission of the American Physical Society.)

109. Conception and Development of the Second Life® Embryo Physics Course (2013)[907]

For a few years, we ran an Embryo Physics Course online, with guest speakers[908]. It was run in Second Life®, an online virtual world, in which each person had their own avatar. Once a week we met and exchanged ideas. The course actually was good for credit as a seminar course for graduate students at Wayne State University.

Figure 203. People as avatars attending the Second Life® Embryo Physics Course[907]. (Open access.)

I was an adjunct at Wayne State University for three years until I had a disagreement with my department head there over how the COVID pandemic was being handled and I was abruptly terminated. Natalie and I both signed the Great Barrington Declaration. We both felt the lockdowns and vaccines mandates were just plain unethical aside from being ineffective and harmful. As a Human Geneticist Natalie had strong reservations about the "safe and effective" narrative being forced on society. Like so many others, we were immediately ostracized for daring to go against the CDC/NIH Big Pharma "science" of the day.

110. Cosmic Embryo #6: Swaying Trees to Produce Wind Energy (2013)[909]

I guess that the world's energy problems were often on my mind, such as solar panels[911], of which we've acquired and installed a few. We had an off-grid cabin that we built ourselves and bought the first LED flashlight that lit up when charged by shaking. One windy

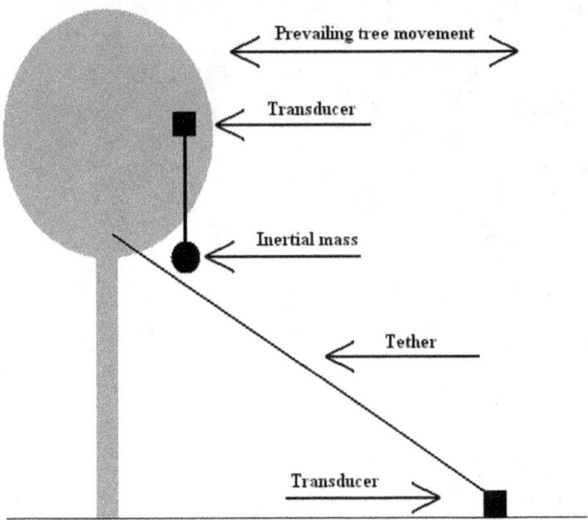

Figure 204.   A tree swaying in the wind, storing wind energy. Experiment to measure wind energy from a tree. (From[910] MDPI open access.)

night at our cabin, when the trees were swaying, I thought about letting the trees shake this flashlight. On looking into the literature, it became obvious that trees absorb most of the wind energy that impinges on them, so the proposal wouldn't work well. As we spent 8 winters at the Gulf Specimen Marine Lab & Aquarium in Panacea, Florida, I became quite aware of the power of hurricanes. This leads to an interesting suggestion that the huge trees in northern Florida might dissipate much of the force of a hurricane, and shouldn't be cut down, as people often do. This idea might be testable by comparing the hurricane damage along forested and deforested coastlines, with data from satellites and records.

111. Cosmic Embryo #7: Is There an Optimum Solution to the Missing Sock Problem? (2014)[912]

One problem with raising kids is that they swipe your socks. I found one solution: buy a huge number of identical socks in a plaid design the kids wouldn't be caught dead wearing[912].

Figure 205.   A pile of my (mostly) identical socks.

112. Soap Bar Skin Scanner for Detection of Early Melanoma (2016)[913]

Many people do their thinking while showering. This proposal combines the detection of position in space with imaging of nevi, 1/100,000[th] of which become melanomas. This cancer is relatively easy to detect on the skin and remove, but is 95% fatal if it metastasizes.

113. Biocommunication: Sign-Mediated Interactions between Cells and Organisms (2016)[914]

This book was an attempt to consolidate communication between cells on up, from Bacteria to primates[677,914]. The next article was my contribution.

114. Cybernetic Embryo (2016)[915]

Rob Stone, a Michigan high school teacher interested in philosophy, and I, took the occasion of this book to try to define levels of purposiveness in biology. I have found myself often skirting around the biosemioticists and their break-off field of code biology. Later

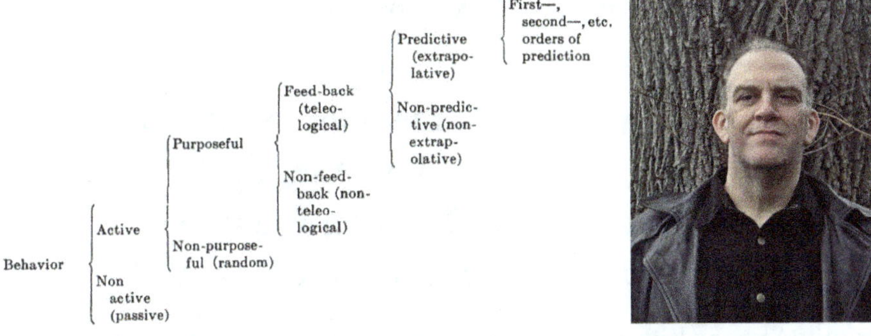

Figure 206. a) An attempt at levels of purposiveness[915]. (With permission of World Scientific Publishing.)   **Figure 206.** b) Rob Stone. (With his permission.)

we phrased this in terms of Janus-faced causality[335,336], after reading Arthur Koestler[916].

115. Morphozoic, Cellular Automata with Nested Neighborhoods as a Metamorphic Representation of Morphogenesis (2016)[917]

Morphozoic[917] is a toy model for morphogenesis based on cell-cell interactions of all ranges, using adaptive neighborhoods[541]. It was a generalization of cellular automata, which I had gotten into with my model for a 2D "snail"[79]. Tom Portegys (Figure 207) works on novel robots.

Figure 207.   Tom Portegys. (With his permission.)

116. Quantifying Mosaic Development: Towards an Evo-devo Postmodern Synthesis of the Evolution of Development via Differentiation Trees of Embryos (2016)[332]

Bradly Alicea (Figure 228) has been a frequent collaborator since he invited me to lecture at Michigan State University in 2013.

In this paper, we extended the idea of the differentiation tree to mosaic embryos, first proposed in[183]. See also[671].

117. Deformation Modes and Structural Response of Diatom Shells (2017)[918]

The buckling patterns of diatom valves can be fascinating[300,565,566], in this paper approached with the help of two mechanical engineers.

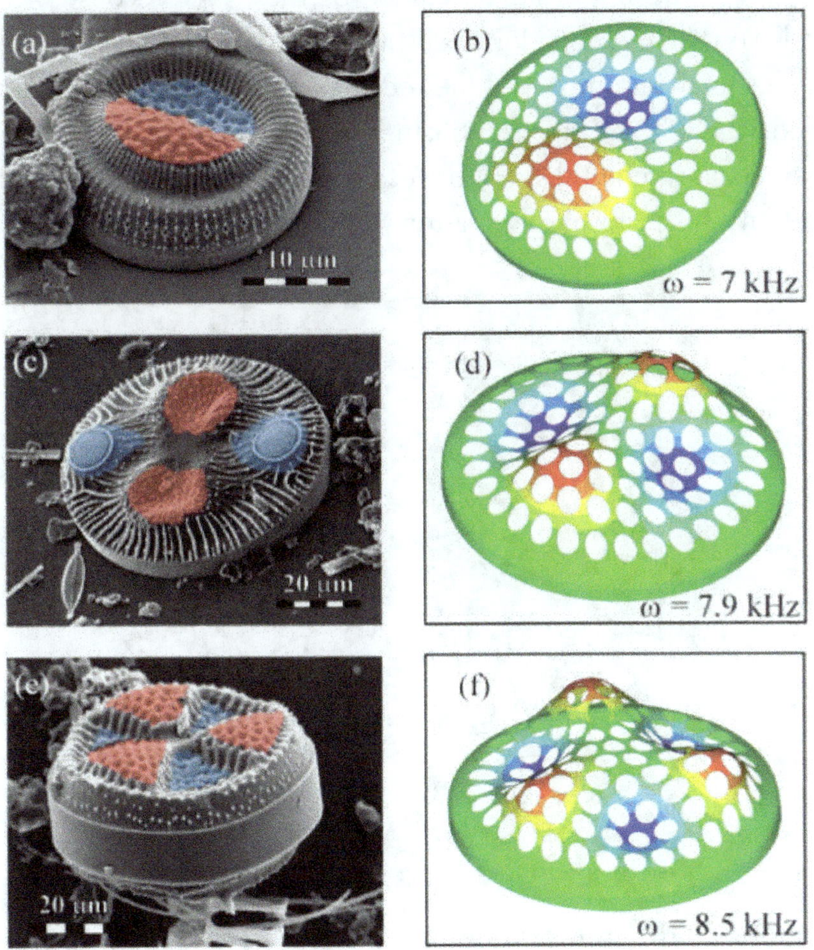

Figure 208. Various modes of buckling of perforated, centric diatom valves[918]. (With permission under the Creative Commons Attribution International License (CC BY 3.0).)

118. Foxels for High Flux, High Resolution Computed Tomography (FoxelCT) using Broad X-ray Focal Spots: Theory and Two-dimensional Fan Beam Examples (2017)[919]

It is generally assumed in the development of X-ray computed tomography algorithms, that the ray width equals the detector pixel width. My graduate student Elzbieta Mazur actually used this constraint in a positive manner, by tilting the reconstruction pixels so that their shadows fell precisely on the detector pixels, which somewhat improved the reconstructed images[920]. This presaged RICT[921] and the pointilism approach to image processing, in which, as in pointillism[922], the exact shape, size, orientation, and position of a pixel hardly matter when an image is viewed from

Figure 209.   Even with a 15° reverse fan beam, foxels reliably reconstruct the center of the Lena test pattern[919]. (With permission under the terms of the Creative Commons Attribution License.)

a far enough distance[201]. The exact position and shape of a pixel could be varied. She and her huband Andrzej escaped Poland before Solidarity came to power, and did PhDs in my lab, becoming medical physicists afterwards. An amusing episode occurred when young Kurt Luchka[923], who also became a medical physicist[924], came to Natalie embarrassed that as a speaker of Ukrainian, he could understand their Polish, when they described what they would do to one another when they got home. Kurt and I also spent a weekend taking turns sleeping in the lab to change the video tapes good for 8 hours each for time-lapse of an axolotl egg, which Natalie concluded had been dead the whole time.

Glen Colquhuon (Figure 52), a retired programmer and I, introduced the idea of Foxels (focal spot pixels)[220], in which we considered what we called the "reverse cone beam" from each point on the x-ray source to the detector pixels. Graham Alvare (Figure 210), a medical student then (now a practicing physician), implemented this and got some excellent reconstructions[919]. Foxels suggest that our planned use of parallel polycapillary x-ray sources for 3D breast imaging, which come in beams much wider than detector pixels, should be deconvolutable.

Figure 210. Graham Alvare. (UC.)

119. Computational, Theoretical, and Experimental Approaches to Morphogenesis (2018)[925]

This was a special issue of the journal *BioSystems* that I helped edit, and just contributed to its editorial[925] (cf.[926]). Abir Igamberdiev (Figure 211), as Editor-in-Chief of *BioSystems*, has made that journal central to discussions of the differentiation wave hypothesis. It is slowly being considered, as for butterfly spot patterns by Joji Otaki [211]. Abir is a theoretical biologist and plant physiologist.

Figure 211.   Abir (Andrei) Igamberdiev. (From[927] under the Creative Commons Attribution-Share Alike 4.0 International license.)

120. Habitability of the Universe Before Earth (2018)[60]

We had a puzzle: if Earth is only 1/3 of the age of the Universe, then maybe creatures like us could have evolved twice before, and if still around, be much more advanced. An origin of life before Earth existed seemed plausible[928]. A meeting on origin of life at Carnegie Institution, Washington, DC, provided an opportunity to

Figure 212.   Alexei Sharov. (With his permission.)

solicit authors at a poster session[929]. A few responded, envigorating my career in astrobiology.

In addition, Alexei Sharov (Figure 212) introduced me to the field of biosemiotics[930] and his hypothesis of the Coenzyme World at the origin of life[931].

121. Why was the Transition from Microbial Pokaryotes to Eukaryotic Organisms a Cosmic Gigayear Event? (2018)[932]

George Mikhailovsky (Figure 43) and I wondered why Earth went 2 billion years before prokaryotes gave rise to eukaryotes. We considered all possibilities we could think of, and decided that this duration could have been a biological phenomenon. We did a crude simulation, concluding that accumulation of genes and horizontal gene transform could explain the delay (Figure 213). This suggestion requires further critical work. See also[934,935]. In general, we seem to be the only ones to question why some stages of evolution took so long.

Figure 213.   A schematic view of the delay from LUCA to LECA[933]. (With permission of Elsevier.)

122. Cosmic Embryo #8: What's Swimming On/In My Eyes? (2018)[936]

Whenever I looked at a clear, blue sky, I saw shadowed onto my retinas tiny dots that seemed to be swimming across my field of view. I never figured out what they were. They weren't simply pulled by gravity. I no longer see them, now that I've had my cataracts replaced by artificial, tiny lenses. This may imply that these motes swam on rather than in my eyes.

Chapter 7
2020s

123. Steps of Silicic Acid Transformation to Silica Frustules: Main Hypothesis and Discoveries (2020)[937]

Vadim Annenkov (Figure 214) has devoted much of his effort to molecular dynamics simulations of diatom pattern formation. In this review, we considered the efforts to understand this process. See also[291,293,938].

Vadim has also taken on preserving the best videos of diatoms in the journal *Limnology and Freshwater Biology*, such as that

Figure 214.   Vadim Annenkov in 2012[945]. (With his permission.)

Figure 215.   Thomas Harbich. (With his permission.)

of Jeremy Pickett-Heaps[939,940]. Thomas Harbich (Figure 215) is a retired engineer who has posted many beautiful web pages on diatom behavior[941,942] and contributed to the mathematical modelling of diatoms[283,867,943,944]. He is leading an effort to summarize diatom research results since Pickett-Heaps' movie[939].

124. Editorial: Symbiogenesis and Progressive Evolution (2021)[935]

The notion of progressive evolution goes back to the late 1800s, and is just recently discussed without the overlay of Victorian ideas of progress. George Mikhailovsky (Figure 43) and I contributed a couple of articles to this special issue[933,934].

125. Diatom Pore Arrays' Periodicities and Symmetries in the Euclidean Plane: Nature between Perfection and Imperfection (2021)[946]

Pores in diatom silica come in a huge variety of shapes, arrangements, and often have subpores. They are presumed to be routes

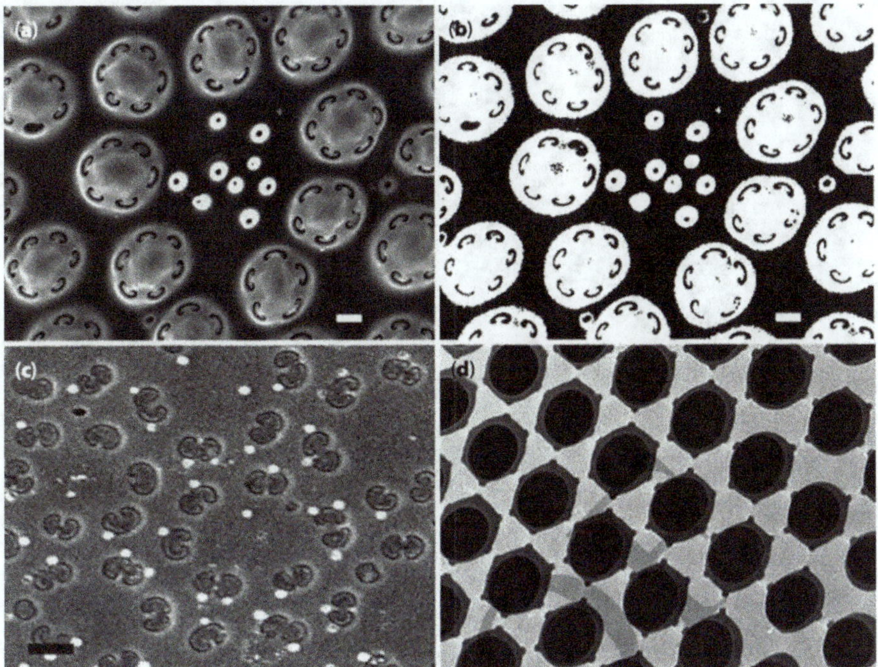

Figure 216.   Some pore varieties in diatoms[946]. (With permission of Wiley-Scrivener.)

Figure 217.   Ghobara and his bride Mia, in Egypt. (With his permission.)

Figure 218.   Mary Ann Tiffany and me in San Diego.

for nutrients and waste products to get past the silica walls. They are involved in valve mechanics[918]. However, their optical properties have proven to be of importance to diatom nanotechnology and perhaps to the diatoms themselves[947]. Some may make great wallpaper.

Two of the coauthors had nonstandard careers. M. Ghobara was a botanist who became a physicist. Mary Ann Tiffany got her PhD as a grandmother, and produced some of the best SEMs (Scanning Electron Micrographs) of diatoms. See also[948]. Some of her SEMs are preserved in a stone bench a GSML.

126. Comparison of Photon-counting and Flat-panel Digital Mammography for the Purpose of 3D Imaging using a Novel Image Processing Method (2021)[430]

As most women (about 7/8) don't get breast cancer, it is important to keep the imaging dose to a minimum, so that the probability of

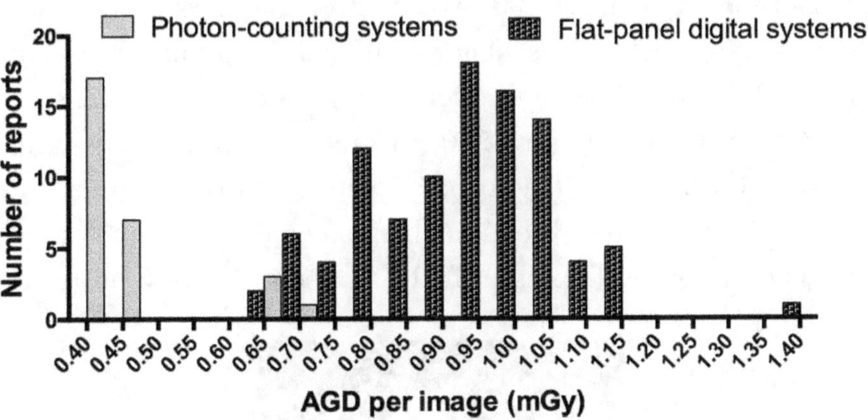

Figure 219. Photon-counting requires less dose[430]. (With permission from Oxford University Press.)

the X-rays causing cancer is insignificant. At the time Tony Svahn, a medical physicist in Sweden, found that for ordinary mammography, photon-counting detectors required less dose. X-ray phase methods[949] and RBRCT promise even lower dose.

127. Terraforming Mars (2021)[950]

This is an example of a book I helped edit, but did not write a chapter. I've supported the possible use of diatoms in providing O_2 on Mars[951,952]. The editor, Martin Beech, has succeeded me as Astrobiology Series Editor.

128. Diatom Triboacoustics (2021)[867]

Sometimes a negative result may be important, but is rarely reported in the scientific literature. I fancied that because the motion of diatoms is so jittery[261], we ought to be able to hear them move. No luck. Even in a professionally quieted room, their motion was silent[867]. Florian Zischka was a graduate student of Ille Gebeshuber (Figure 190). We are now considering a different approach, in which the visual jittery motion (apparently Brownian plus drift) is turned into a frequency range audible to humans.

Can Sabuncu (Figure 220), a graduate student and then postdoc of Ali Beskok (Figure 221), and I determined that diatom motility until the direction changed could be characterized as a 1D random walk plus drift. I had been a visiting professor

Figure 220.   Can Sabuncu[954]. (With his permission.)

Figure 221.   Ali Beskok[955]. (With permission of the Hunt Institute for Engineering and Humanity.)

at Old Dominion University and later worked with them at Southern Methodist University because Ali is a world expert on nanofluidics[953], which seemed to describe the motion of raphe fibers inside raphes.

129. Employing Newly Developed Plastic Bubble Wrap Technique for Biofuel Production from Diatoms Cultivated in Discarded Plastic Waste Bubble Wraps Collected from Bubble Wrap Industry (2022)[852]

When you're retired and without a lab or equipment, if you want to do an experiment, some improvisation is necessary. I was concerned which plastic might retain water and allow gas exchange, so I just used air pillows, the larger form of bubble wrap. This LDPE (Low Density Polyethylene)[956] proved to be the right choice, resulting in this paper with Vandana Vinayak and her crew[852].

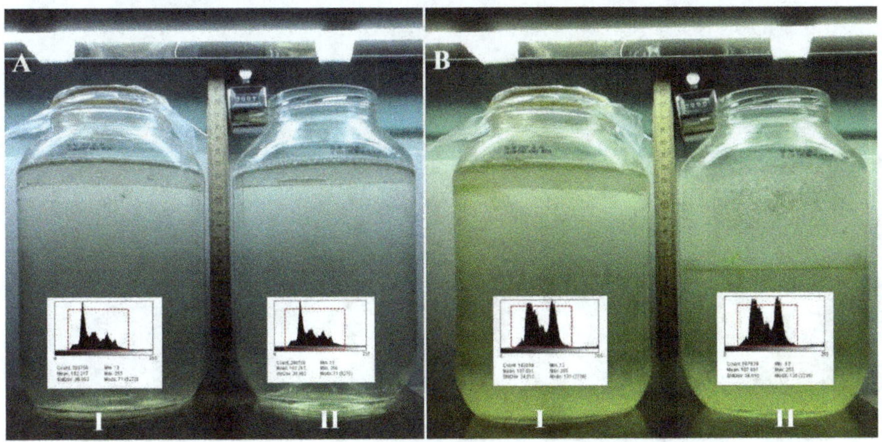

Figure 222. Jars of pond water at collection and 50 days later, in a home experiment. The one on the left was covered with low density polyethylene from an air pillow. There was no water loss. The algae grew equally well with the exposed bottle, suggesting gas exchange (O_2 and CO_2) was not hindered by this plastic, but water was. (With permission of Elsevier.)

This work was preceded by a book on algal fuels[957]. A book on diatom culture follows[297]. It has also been taken up by Cliff Mertz[858], and if scalable to vast farms (Bubble Farming) has the potential of providing contamination-free and zero carbon footprint biofuels and possibly a new Green Revolution[849,958,959], as some species are edible. However, we have had no indication from governments or industry that they have any interest in a solution to their touted "climate crisis" that allows gasoline engines to persist and has a near zero carbon footprint. In the meantime, the destruction of the Malaysian rainforest proceeds apace to provide nickel batteries for electric vehicles[960].

130. Types of X-ray Techniques for Diatom Research (2022)[961]

Computed tomography (CT) is now everywhere, and should be taught in linear algebra and maybe replace calculus as the gatekeeper for college entrance. I have long thought of doing a high school book on CT. As my approach to CT, "CT brush"[962] (Figure 223) has now been declared a "serious" computer game[963],

Figure 223.   "A mockup of what the screen might look like in a CancerZap! video game. The player shoots x-ray photons at a scene that is rotating, with one object (the Martian eye) in this case representing the "bad guy" tumor. The gun's lateral motion would correspond to a fan beam. As in CT Brush, the objects would only become visible as they were shot at, accumulating x-ray dose"[962]. (With permission via Creative Commons CC BY 4.0.)

that may be where to begin interesting high school kids in CT. I won't start to list its many applications[213], except for one, that particularly pleased me: 3D imaging of the solar corona[888]. In this chapter[961] of[294], we discussed its application to the 3D structure of diatoms[961].

131. Conflicting Models for the Origin of Life (2023)[964]

This book[964] was my first full eye opener to the plethora of models for the origin of life[965]. Stoyan Smoukov is one of the chemical engineers who discovered shaped droplets in 2014.

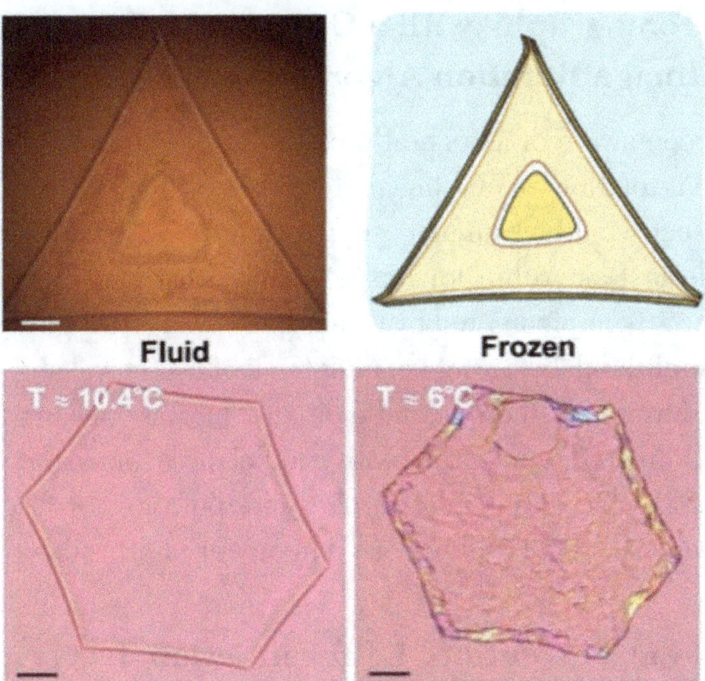

Figure 224.   "Light micrographs showing the thicker border, and the thin film in the middle, of shaped droplets. In the triangle a sizeable thin patch has formed inside, whereas in the hexagon it has stretched all the way. Sometimes in these shapes, because the middle film is so thin, they puncture before they can stretch all the way, creating holes. The second triangle is a sketch of the first. Scale bars = 20 μm." (From[73] with permission of Wiley-Scrivener.)

Figure 225.   Stoyan Smoukov[1].

132. RICT: Rotating Image Computed Tomography with a One-to-One Reversible Image Rotation Algorithm (2023)[921]

I had been aware of a general problem in image processing that repeated rotation of an image introduced small interpolation errors that would accumulate, destroying the image, I came up with the first algorithm for image rotation that does not do this (Figure 226), implemented by Chengxiang Wang[921] (Figure 80). It was complicated, involving keeping track of each pixel and where it went during each incremental rotation. This was done in the context of computed tomography with parallel projections, but should be generalizable. But given the many interpolation approaches, it is unlikely that RICT will be seized upon nor improved.

133. Macroevolution, Differentiation Trees, and the Growth of Coding Systems (2023)[338]

I redefined macroevolution in terms of differentiation trees[183]: macroevolution occurs when the topology of the differentiation

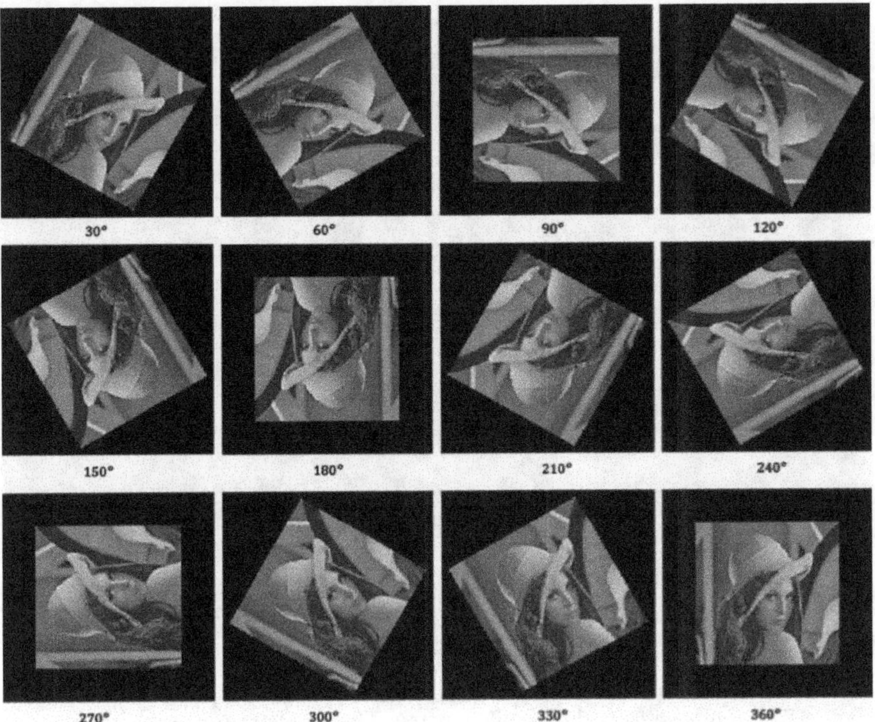

Figure 226. Lena rotated by RICT[921]. (With permission of IOS Press.)

tree changes; microevolution occurs when it does not. Abir Igamberdiev (Figure 211), Editor-in-Chief of *BioSystems*[966], has worked with me in reintroducing this concept and bringing it up to date.

With Bradly Alicea (Figure 228), I've written many papers on lineage trees, other aspects of morphogenesis, diatom motility and other topics[262,264,332,625,671,672,917,925,967–971]. Much of this productivity stemmed from his online student collaborators in DevoWorm.

The late Donald Williamson[972,973] (Figure 229) was a zoologist who, before the Human Genome Project, hypothesized that many organisms were basically concatenations of two disparate species[974–976]. After extensive correspondence with him, I tried to

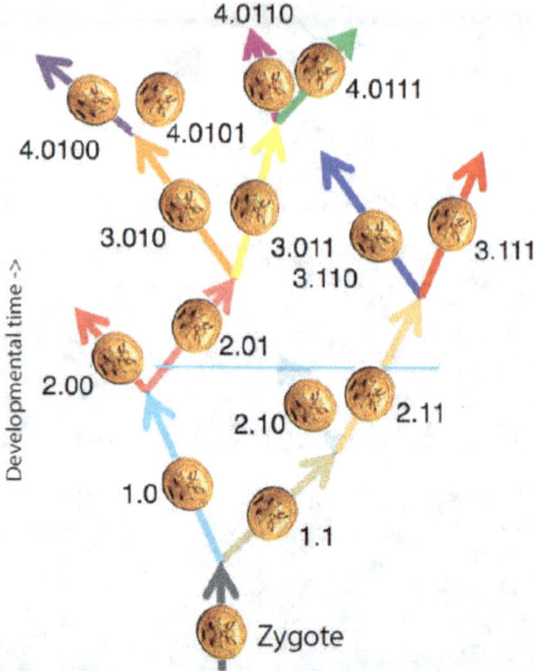

Figure 227. A schematic differentiation tree. The numbers are the differentiation codes. The colors of the arrows indicate that each cell type is different, The Roman Janus-faced coins suggest the global/local interactions at each step of differentiation. (From[149] with permission of World Scientific Publishing.)

Figure 228. Bradly Alicea. (With his permission.)

Figure 229. Donald Williamson (1922–2016)[972]. (With permission of Oxford University Press.)

categorize various metamorphoses in terms of concatenated differentiation trees[183], discussed further in[338,968]. His idea deserves a modern genomics approach. Criticisms of it occurred before any genomes were sequenced. Now the many sequenced organisms should permit analysis of his proposed genome mergers.

134. NoRCEL's Engagement in Africa: The AstroScience Exploration Network (ASEN) (2023)[977]

I attended a course on Africa at the University of Oregon with 50 Africans from the newly independent countries during the 1960s, partying with them, and so always had a warm spot for Africa, meeting and attempting collaborations with African epidemiologists, once to help out with halting the AIDS epidemic in Afghanistan. Thus, I joined NoRCEL for a while, suggesting that they emphasize the less expensive theoretical astrobiology. One African, Jackson Achankunju, did join an effort at my suggestion[978].

135. Embodied Cognitive Morphogenesis as a Route to Intelligent Systems (2023)[968]

Edmund Sinnott[980] (Figure 231) hypothesized that there was an intimate relationship between morphogenesis and mental phenomena[981–985]. He seems to have been ignored. Despite rising in

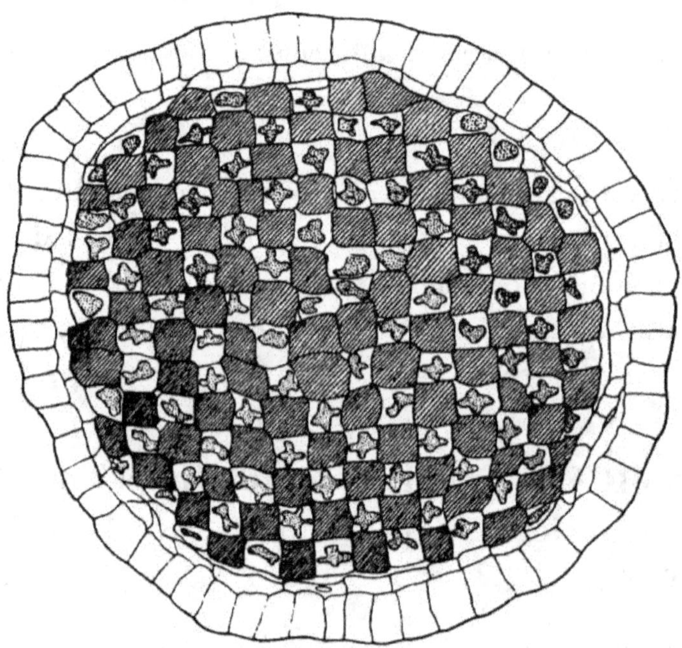

Figure 230.   An unexplained checkerboard phenomenon from Goebel kept alive by Sinnott[979]. "Section through developing liverwort sporangium showing differentiation of alternating spores and elater cells. (From Goebel.)" In public domain.

Figure 231. Edmund Sinnott (1888–1968)[980]. (Public domain.)

Figure 232. Steve Levin. (With his permission.)

the ranks of science, this aspect of his work seems to have been forgotten[980]. I have discussed his viewpoint in[183]. I was also heavily influenced by his masterful book on plant morphogenesis[979]. (Cf.[262].)

136. Is it a Janus-faced World after all? Physics is Not Reductionist (2023)[337]

Every once in a while, I indulge as an amateur in fundamental physics. Bashir Ahmad (Figure 149), who was a key figure in Books With Wings in Herat, Afghanistan, always wanted to be a physicist, and received an unprecedented full scholarship from the University of Manitoba to do an undergraduate degree in Physics. While physics is usually thought of as reductionist, and biology as requiring many levels of explanation, in this paper we showed that they aren't so different. I have also indulged in proposing a tensegrity structure for the atomic nucleus, which is still a conundrum[73,986]. I learned much about tensegrity from Stephen Levin (Figure 232) when he attended a conference on origin of life with me at the Carnegie Institution of Washington, DC. I reviewed tensegrity and tensegrity art in an appendix to[987].

137. Origin of Life via Archaea: Shaped Droplets to Archaea First, With a Compendium of Archaea Micrographs (2024)[73]

The first chapter of this book explains what it's all about:

> "In 2014 a remarkable, counterintuitive discovery was made by chemical engineers that oil droplets in water, if cooled slowly, departed from their expected spherical shape (like oil drops in shaken Italian salad dressing), instead forming flat polygons with sharp corners. On coming across this work during my eclectic reading as a theoretical biologist, I grew very excited, contacted the authors, and proposed that they had found a missing link in attempts to explain the origin of life, a paper on which some of them joined me[987]."

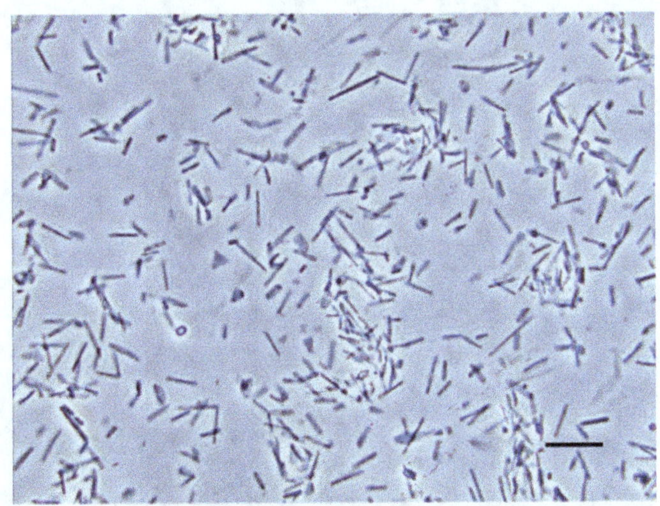

Figure 233. One triangular Archaea amongst many rod-shaped phenotypes. Sometimes finding the polygonal cells is like finding Waldo[73]. (From[988], open access.)

In this book, I showed that 41% of extant Archaea species have flat, polygonal phenotypes (Figure 233), and apparently no Bacteria, suggesting that Archaea cells such as these may have been the first living creatures on Earth. I also mention many other hypotheses, in a chapter called *A Survey of the Battlefield for the Origin of Life*, showing there is no consensus, and the problem of origin of life remains quite open. My earlier interests in the origin of life go back to work with Bruce Damer [719,989] (Figure 234), who has been working with David Deamer[990] (Figure 236), an inspiration to this, my book on the Lipid World hypothesis for the origin of life[73]. I started with Bruce and Thomas Barbalet (Figure 235) trying to simulate open evolution, the ability to generate ever more complex organisms[719].

During this work, I was also able to explain the polygonal shapes of some Archaea based on the energetics of their external, membrane embedded, protein S-layers[993].

Figure 234.   Bruce Damer[991]. (Under the Creative Commons Attribution-Share Alike 3.0 Unported license.)

Figure 235.   Thomas Barbalet, an artificial life simulator[719,992]. (With his permission.)

In this book, I introduced the contrarian notion that much of the origin of life could have occurred at thermodynamic equilibrium, without requiring an energy source, given a sufficiently large ensemble of vesicles or other "containers"[994], and a physiochemical

Figure 236.   David Deamer[990], NASA. (Public domain.)

model for the first membrane peptides[181]. I also supported the as yet unproven notion that autocatalysis could arise spontaneously[995].

138. The Chromolinker Hypothesis: Are Eukaryotic Genomes also circular? (2024)[996]

Most Archaea and Bacteria have circular genomes, whereas most Eukaryotes are presumed to have individual, separate straight chromosomes. As Eukaryotes are hypothesized to have been formed by some kind of fusion of an Archaea and a bacterium, this anomaly needs to be accounted for, amongst others. Over the past century there have been numerous reports, pro and con, that like most prokaryotic genomes, the eukaryotic genome is circular, including a report by Andy Maniotis that the whole genome may be pulled out of some cells[997]. The purpose of this article is to bring this contradictory literature to the attention of young scientists who may wish to resolve this conundrum. Natalie speculated on the possibility of chromolinkers and their mechanism in mitosis and meiosis in[149]. In this paper I attempt to review the hints, pro and con, suggesting that with modern approaches, it is time to resolve this question. It might have a major impact on our views of genetics and evolution.

139. The Molecular Basis of Differentiation Wave Activity in Embryogenesis (2024)[675]

We've attempted to guesstimate the relationship between the genetic code and differentiation waves[675]. This attempt is contrary to the "higher than genome" approach of Philip Ball, albeit including it, who was apparently unaware of differentiation waves and their possible ability to trigger specific genes as they traverse an embryo to explain the many phenotypes in multicellular organisms, despite the genome being the same in most cells[998]. Our approach could use rigorous testing, but because it contradicts the molecular developmental approach that genes are the primary source of development, I expect only posthumous testing[336], perhaps by the next generation.

140. Electronic Steering of X-rays (2024)[257]

It has been an old dream of steering X-rays, going back to 1950[999]. Here Hussain Ather (Figure 238) and I give a method of doing so for the future (Figure 237).

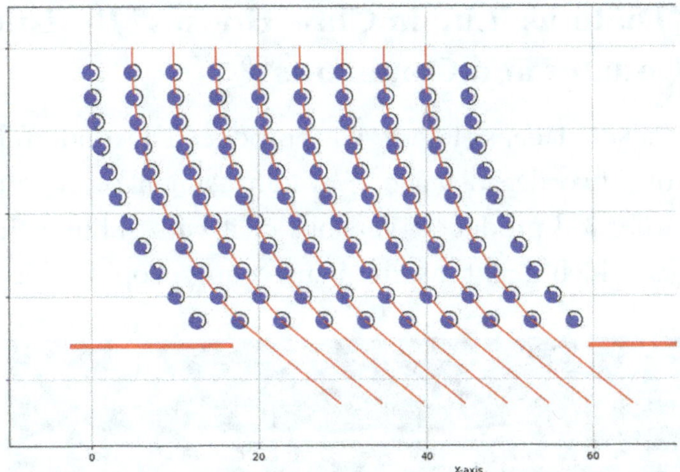

Figure 237. 11 rows of Janus spheres undergo Bragg reflection of a beam of parallel X-rays from a parallel polycapillary source. At the bottom is a movable collimator to block most scattered x-rays.

Figure 238. Hussain Ather. (With his permission.)

141. "Diatoms: Life in Glass Houses" Revisited: Updates and Comments[939]

Jeremy Pickett-Heaps (Figure 240) produced a wonderful movie by this title two decades ago[1001]. Vadim Annenkov (Figure 214) has spearheaded producing versions of it with subtitles in many languages, which might inspire many young people worldwide to

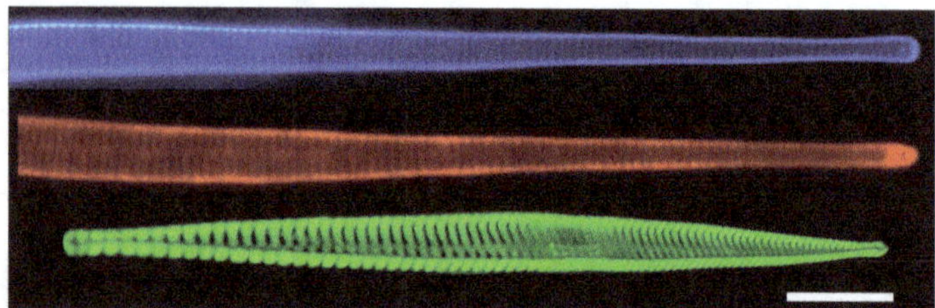

Figure 239.   "Fluorescence and confocal (bottom) images of different *Ulnaria* sp. valves after culturing with the addition of dyes (blue: Q-N2; red: Rhod-N3H; and green: NBD-N2; refer to table 1 for references). Developed over the past 20 years, these dyes are incorporated into newly formed frustule parts, enabling the expansion of the range for targeted exploration in experiments requiring specific wavelengths. Experimental details for the doping procedure are available in[1000]. Scale bar represents $10\,\mu$m". (Open access[939].)

Figure 240.   Jeremy Pickett-Heaps (1940–2021). (With permission of the photographer, Stanley Cohn.)

do research on diatoms. This publication updates many aspects of diatom research illustrated in the movie.

142. Steerable Ray-by-Ray X-Ray Computed Tomography: A Plan to End Breast Cancer (2025)[136]

I have focused on breast cancer since 1977[201,477,775,1002]. It has lagged decades behind the advances in CT, my experience being that women raise money via "runs for the cure" for men to do the research, who are in no rush to end their largess, rather than women doing the harder work of educating themselves to do the research themselves. There is a new polycapillary X-ray source, perfected over the past two decades, that might solve the problem[219] (cf.[1003]). While all cancers are biochemically complex, my approach is "search and destroy" without biopsy, which sometimes

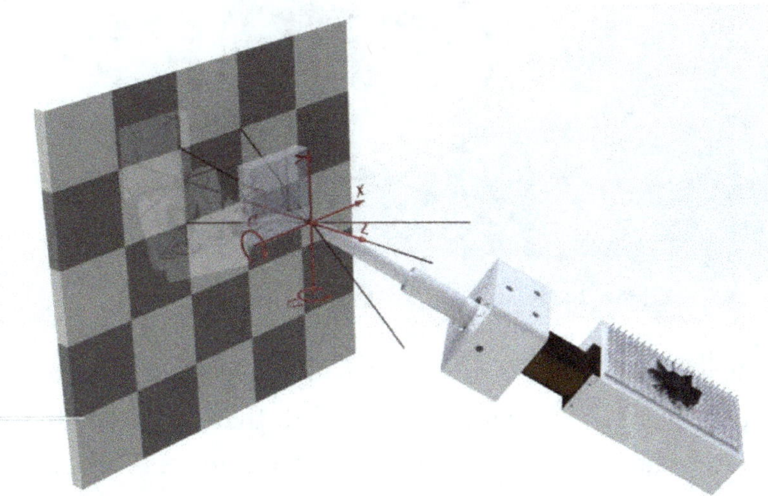

Figure 241. A polycapillary X-ray source aimed at an angle into the top breast compression plate. By Marcelo Valderrey[136] (Figure 243), forthcoming. (With permission of Martin Scrivener of Wiley-Scrivener.) Breast pathologist Tibor Tot has helped keep this work down to Earth.

Figure 242.   A CAD (Computer Aided Design) for a robot controlled polycapillary X-ray source for 3D mammography and a second robot to control the breast compression plates. By Marcelo E. Valderrey[136] (Figure 243), forthcoming. (With permission of Martin Scrivener of Wiley-Scrivener.)

Figure 243.   Marcelo Valderrey. (With his permission.)

spreads a cancer. This might be possible, as I have worked with epidemiologists[1004–1007] and extrapolated that, if we could find 100% of breast tumors of size up to 4 millimeters in diameter, and destroy them, we might achieve a greater than 99% cure rate of focal breast cancers[772,1008,1009] (Figure 244). These polycapillary sources may make steered beam x-ray imaging[962,1010] practical.

This approach requires ray-by-ray CT (RBYRCT), to minimize the dose, which has never been attempted, and thus new CT algorithms. It also requires development of a new apparatus, which we hope will cost less than current mammography machines, and perhaps be adaptable to them. The idea of ablating them comes from Mark Torchia's research on laser ablation of tumors (Figure 245) and his comment: "If you can find them,

Figure 244. Vincent Vinh-Hung. (With his permission.)

Figure 245. Mark Torchia. (With his permission.)

I can ablate them". (His method plus MRI, due to Paul Lauterbur, Figure 92, probably saved the life of one of my sons, who developed a brain tumor just after a "safe and effective" COVID vax injection.) Curiously, steering of the x-rays is accomplished by a robot carrying polycapillary optics or perhaps in the future electronically by Bragg reflecting Janus spheres. Bragg reflection with much longer wavelength is being used to quantitate ocean currents[1011]. I recall one night spending a long time watching the oscillating reflection of the Moon from small water waves under a bridge over a canal in Denmark, after reading papers about the statistics of glint, and then buying a colorful sweater which was buttoned on the "wrong side", as if anyone might notice my *faux pas* of a button side reserved in Denmark for females. I still have the tiny wooden shoes I bought for my then young daughter Lana.

143. Conclusion

I never thought of myself as a team player: I lost my leather baseball catcher's mitt as a kid which conveniently allowed me to drop out of Little League, and at most joined scientific societies for brief periods, hardly ever going to two meetings by the same group. I thus lost all opportunities to be a climber, and only received one minor award in my lifetime[10]. This might have been due to a distasteful experience: the newly elected President of the Society of Developmental Biology, Paul B. Green, an excellent theoretical botanist, asked me to be Secretary, and then embarrassingly had to tell me that the former executive had already chosen someone. After having attended for a few years, I never did again. Nevertheless, I have collaborated with many people. Many of my collaborators did not make it in the grant grabbing world of academia[629]. I just squeaked through, via multiple agencies, national and local[1012], and diverse projects and departments. I only did what I wanted to, from the trivial to major questions. But grant writing often robbed me of months of research, and was often fruitless.

Some seemed to run with my work. While I begrudged that, as I've honestly written above, it is the way science works: everything I've done is public, and thus available to everyone. One can think of it as annoying, or as the honor of having one's work built upon, be it credited by citation or not. I too have leaned on far more people and their work than I could possibly mention here. For example, my first monograph book[183] has over 6000 references, many with multiple coauthors. The number of references in[73] is almost as high. The tens of thousands of scientists on whose work I have built couldn't possibly all be acknowledged here.

As I've shown, being a theoretical biologist is generally a lonely path. Nevertheless, the encouragement of even one person makes it all worthwhile:

"Pieter [Nieuwkoop] said. "If you have done the evolution and genetic research as well as you have done the embryological then you have done it well indeed, at least in so far as one can learn a field and then speculate without reservation about virtually everything in it!" Pieter chuckled then. " "You do not lack, what is that wonderful Jewish word?"

"Chutzpah".

Ah yes, chutzpah, you have no lack of chutzpah, that much is certain."

"But what if I'm right?" Dick asked.

"If you are right," Pieter replied, "then you will have succeeded in unifying development, genetics, and evolution. This would be a great accomplishment. But your ideas are so far beyond the available data that you will not live long enough to know if you're right. There are generations of work here, hundreds of Ph.D. projects, with whole careers worth of work here, proving you are right, or possibly that you are completely wrong. You also assume you are right, and that may be assuming much. I personally still reserve judgment on how right you are."

Now it was Dick's turn to laugh. "If my ideas are properly tested, then even if I am wrong, that's all I want because much new research will be generated and we'll accumulate a lot of new knowledge. What I want to do is provide a fresh perspective on the old problems."

"It definitely is a fresh perspective," Pieter agreed. "So fresh some people may never speak to you again!"[183].

As is obvious I am an avid, if slow, reader (Figure 11). What is not obvious is that it was about everything nonfiction. (I can't follow novel story lines.) For instance, I collected over 2000 books, mostly on my world travels which usually included going to local used book stores. The books occupied over 40 bookshelves, including one of biographies and autobiographies of scientists, which my students helped assemble, a long, hard job before powered screwdrivers were available. I owned most of the many books by Darwin and other rare books, such as Ernst Haeckel's (Figure 246)

Figure 246.   Ernst Haeckel (1834–1919). (Public domain.)

Kunstformen der Natur (*Art Forms of Nature*[1013]), which I donated to the medical rare book room at the University of Manitoba, and read his posthumously published love letters[1014], which showed him torn between keeping his reputation or happiness. She died suddenly, resolving his problem.

Still, with those books I kept, it was not enough of a reduction for trailer life. *En route* to give a lecture in Los Angeles at UCLA, pulling our trailer/home, Natalie and I took a "short cut" over a mountain road from Death Valley and burned out all of our truck's brakes. We just made it down alive, as we couldn't stop anywhere along the >20 mile descent, saved only by the trailer's brakes. The repair costs used up our US$1700 honorarium. Despite downsizing, for our 5 years of traveling the USA that way, I had to dispose of most of my remaining books, to further reduce weight for our upcoming trip from Vancouver over mountains to Calgary. In addition, I had had two rooms in my lab crammed with 50,000 reprints in numerous filing cabinets, that were always lapsed in refiling (the days of basement housed professors like me affording a secretary on grants long over). These subsequently were replaced by an additional almost 75,000 reprints that fit digitally inside my growing computer memory. This raises the curious question from a quantum mechanics point of view, as to whether the weight of my sequentially purchased computers goes up or down when the latter digital reprints are added? My quantum mechanics knowledge is elementary, so I'm stuck not knowing. Now I own mostly Kindle books and a laptop computer with 4TB internal memory.

In retrospect, while I was never any good as an artist (Figure 247), never rising above a colorful sketch of a fancy guppy, art figures frequently in my science. Fortunately, after I was forcefully retired, I joined Joseph Seckbach[826,1015] (Figure 248) and

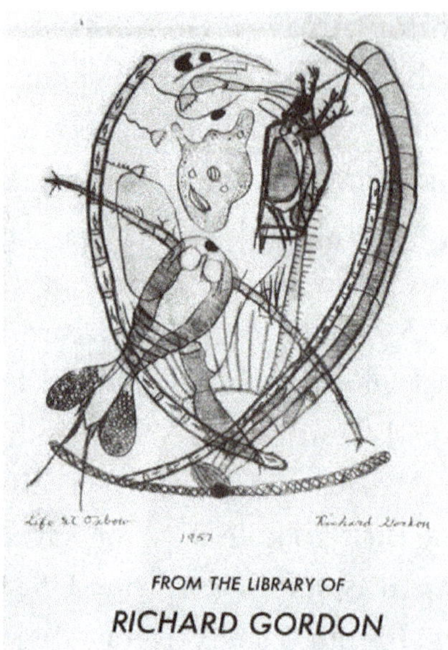

Figure 247. Life at the art colony in Oxbow, Michigan[1022], my lithograph at age 14, used as a bookplate.

Figure 248. Joseph Seckbach in 2012[1023]. (Under the Creative Commons Attribution-Share Alike 3.0 Unported license.)

learned how to edit books and finally came across a great publisher, Martin Scrivener (Figure 249), who made me and Joseph series editors (and editors and for me, a monograph author) for books on diatoms[87,150,293–298,860,891,1016] ("eine lange Krankheit", per Heimie

Figure 249.   Martin Scrivener of Wiley-Scrivener. (With his permission.)

Gesser) and astrobiology[73,157,930,950,964,1017–1020]. I quit those editorial roles after 16 years, needing time to concentrate on my research.

I'm actually amazed that, in pulling this book together, how much of the universe and social affairs I've nevertheless managed to tackle, despite rebuffs, with many uncompleted papers, rejected papers, unreviewed preprints, cobweb sites and mostly rejected grants that I haven't bothered the reader with, papers that didn't make it past the idea stage, and the vast number of men and women around the world who influenced what I did besides those mentioned here. Although I sometimes claim to have specialized in embryogenesis and CT algorithms, I've done much more. Sure, some got over my head (compensated sometimes by people whose math is over my head, Figure 250), and I either veered off or recruited others better equipped. I even find that some of my own early papers are now over my head! Disuse of math or languages, human or computer, makes one rusty. There are many loose ends,

Figure 250. Janice Pappas, lead editor and author on diatom mathematics and art books[87,295,891] and joint papers[299,300,566], now series editor for diatoms. (With her permission.)

some tantalizing unsolved problems, which the reader is welcome to tackle. As the song goes, "I did it my way"[1021], without regrets, except for the shortness of life. I've supervised 53 undergraduate theses and had 3 Masters students, 5 PhD students, 5 postdocs, and 6 visiting professors. Thus, mentoring these and many other much younger collaborators in a way extends my life.

Acknowledgements

Readers who suggested some improvements: Natalie K. Gordon, George Mikhailovsky, Martin Scrivener, plus the many people mentioned and pictured here, where possible, who have taught me much. Thanks to Sook Cheng Lim for her patience as my editor at World Scientific Prublishing. Anyone I've left out here should attribute it to my limited memory or capacity.

There are many people whom I met and/or greatly influenced me, such as restauranteur, cattle farmer, fisherman and Reeve Stanley Asham (1947–2024), code biologist Marcello Barbieri, cytologist Guenter Albrecht-Buehler, cyberneticist William Ross Ashby (1903–1972), breast epidemiologist John C. Bailar III (1932–2016), restauranteur Hong-Jun (Frank) Chen, philosopher Daniel Dennett (1942–2024), molecular biologist Victor Fried, super-resolution pioneer Roy Frieden, botanist Brain W. Fristensky, brider and conservation officer Harry Harris, tensegrity of cytoskeleton investigator Donald E. Ingber, chemist Hyman D. Gesser, (1929–2014), physicist Gerhard Grössing (1957–2019), geologist and astrobiologist Robert Hazen, cell biologist Shinya Inoué (1921–2019), genetic determinism contrarian Evelyn Fox Keller (1936–2023), calcium wave researcher Lionel Jaffe (1927–2011), cell nucleus microscopist Sabine Mai, hologram physicist and entrepreneur Stephen McGrew (1925–2021), embryologist Beatrice Mintz (1921–2022), electron microscopist Donald

Parsons (1928–2017), evolution philosopher Michael E. Ruse (1940–2024), dinosaur paleontologist Paul Sereno, smooth muscle physiologist Newman L. Stephens (1926–2019), art and science historian David R. Topper, artist Frans Visscher (1929–2014), and cyberneticist Heinz von Foerster (1911–2002).

References

1. Thomas Young Centre. Stoyan Smoukov, https://thomas youngcentre.org/people/dr-stoyan-smoukov/ (2024).

2. Colburn JG. Judith Georgia Colburn Greensburg, Pennsylvania. *Florida Southern College — Interlachen Yearbook (Lakeland, FL) — Class of 1959*. 1959.

3. Colburn JG. Judith Georgia Colburn Greensburg, Pennsylvania. *Interlachen, the Florida Southern College Yearbook (Lakeland, FL) — Class of 1961*. 1961.

4. Wikipedia. File:Emile Duclaux.jpg, https://commons.wikimedia.org/wiki/File:Emile_Duclaux.jpg (2021).

5. Duclaux E. *Pasteur-- The History of a Mind*. W.B. Saunders Company, 1920.

6. Wikipedia. File:Isaac.Asimov01.jpg, https://en.wikipedia.org/wiki/Isaac_Asimov#/media/File:Isaac.Asimov01.jpg (1959).

7. Nissani M. Ten cheers for interdisciplinarity: The case for interdisciplinary knowledge and research. *The Social Science Journal* 1997; 34: 201–216. DOI: 10.1016/S0362-3319(97)90051-3.

8. University of Manitoba College of Medicine Archives. Thurlbeck, William Michael, https://umlarchives.lib.umanitoba.ca/thurlbeck-william-michael-4 (1977).

9. Gordon R. Careers in theoretical biology. *Carolina Tips* 1993; 56: 9–11.

10. Rh Institute. Grant for Outstanding Contributions to Scholarship and Research in the Interdisciplinary Category, $2,000, (1981).

11. Herbert D. *Mr. Wizard's Science Secrets*. Popular Mechanics Press, 1952.

12. Andrews RC and Voter TW. *All about Dinosaurs*. Random House/Outlet, 1952.

13. Herbert D, Beauchamp WL, Barker RA, Books H and Company PM. *Mr. Wizard's Science Secrets*. Hawthorn Books, 1961.

14. Andrews RC and Voter TW. *All about Dinosaurs*. Random House, 1956.

15. Zekonja. John Sloan [obituary], https://www.ushandball.org/index.php/about/national-champions/108-bios/370-john-sloan (2016).

16. Kanigel R. *The Man Who Knew Infinity: A Life of the Genius Ramanujan*. New York: Washington Square Press, 1991.

17. Cepelewicz J. Math Is Still Catching Up to the Mysterious Genius of Srinivasa Ramanujan, https://www.quantamagazine.org/srinivasa-ramanujan-was-a-genius-math-is-still-catching-up-20241021/?mc_cid=9a924f365f (2024).

18. University of Chicago. Ryerson Astronomical Society, https://blueprint.uchicago.edu/organization/ras (2024).

19. Gordon R and Sharov AA. Dedications. In: Gordon R and Sharov AA (eds) *Habitability of the Universe Before Earth [in series: Astrobiology: Exploring Life on Earth and Beyond, eds Pabulo Henrique Rampelotto, Joseph Seckbach & Richard Gordon]*. Amsterdam: Elsevier B.V., 2018, pp. v-xi.

20. Musgrave JB. *UFO Occupants & Critters: The Patterns in Canada*. New York, NY, USA: Global Communications, 1979.

21. Wikipedia. File:Roy Chapman Andrews 50488r.jpg, https://commons.wikimedia.org/wiki/File:Roy_Chapman_Andrews_50488r.jpg (1920).

22. Wikipedia. Robert Bateman (painter), https://en.wikipedia.org/wiki/Robert_Bateman_(painter) (2024).

23. Newton I. Philosophiæ Naturalis Principia Mathematica, (1687).

24. Wikipedia. Wallace Kirkland, https://en.wikipedia.org/wiki/Wallace_Kirkland (2024).

25. Kirkland W. *The Lure of the Pond*. H. Regnery Company, 1969.

26. Kalyvas S and Wikipedia. File:Phases Venus.jpg, https://commons.wikimedia.org/wiki/File:Phases_Venus.jpg (2005).

27. Wikipedia. University of Chicago Laboratory Schools, https://en.wikipedia.org/wiki/University_of_Chicago_Laboratory_Schools (2024).

28. Rickey DW and Gordon R. Unitron 60 mm Refractor, https://www.cloudynights.com/topic/10804-unitron-60mm-refractor-classic-scope/ (2004).

29. Gordon R and Sinnott RW. A cyclopean telescope. *Sky & Telescope* 1983; 65/66: 355.

30. Gordon R. A critical review of the physics and statistics of condoms and their role in individual versus societal survival of the AIDS epidemic. *Journal of Sex & Marital Therapy* 1989; 15: 5–30.

31. Kowarski I. How Long Does It Take to Get a Ph.D. Degree? Earning a Ph.D. from a U.S. grad school typically requires nearly six years, federal statistics show., https://www.usnews.com/education/best-graduate-schools/articles/2019-08-12/how-long-does-it-take-to-get-a-phd-degree-and-should-you-get-one (2019).

32. Anders E, DuFresne ER, Hayatsu R, DuFresne A, Cavaillé A and Fitch FW. Contaminated meteorite. *Science* 1964; 146: 1157–1161.

33. Rasband W. *NIH Image (ImageJ)*. http://rsb.info.nih.gov/nih-image/: U.S. National Institutes of Health, 2020.

34. Kadonaga JT. In Memoriam: E. Peter Geiduschek, 1928–2022, https://biology.ucsd.edu/about/news/2022/article_050422b.html (2022).

35. Geiduschek EP. E. Peter Geiduschek. *Curr Biol* 2010; 20: R694-R695.

36. Geiduschek EP. Factors controlling reversibility of DNA denaturation. *Journal of Molecular Biology* 1962; 4: 467-&.

37. Wikipedia. Nicolas Rashevsky, https://en.wikipedia.org/wiki/Nicolas_Rashevsky (2024).

38. Rashevsky N. *Mathematical Biophysics*. University of Chicago Press, 1948.

39. Rosen R. Reminiscences of Nicolas Rashevsky, https://en.wikipedia.org/wiki/Nicolas_Rashevsky (1972).

40. Wikipedia. Robert Rosen (biologist), https://en.wikipedia.org/wiki/Robert_Rosen_(biologist) (2024).

41. Kilkenny L. A Day in the Life: Professor Helmut Hirsch, Distinguished Teaching Professor, https://www.albany.edu/main/features/2005/10-05/1hirsch/hirsch.shtml (2005).

42. Easton J. Robert Uretz, former professor and dean, 1924–2018, https://www.uchicagomedicine.org/forefront/biological-sciences-articles/2018/november/robert-uretz-obituary (2018).

43. Uretz RB and Zirkle RE. Disappearance of spindles in sand-dollar blastomeres after ultraviolet irradiation of cytoplasm. *Biol Bull* 1955; 109: 370.

44. Zirkle RE and Bloom W. Irradiation of parts of individual cells. *Science* 1953; 117: 487–493. 1953/05/08. DOI: 10.1126/science.117.3045.487.

45. Drum RW, Gordon R, Bender R and Goel NS. On weakly coupled diatomic oscillators: *Bacillaria's* paradox resolved. *J Phycol* 1971; 7: 13–14.

46. Crewe AV, Wall J and Langmore J. Visibility of single atoms. *Science* 1970; 168: 1338–1340. DOI: 10.1126/science.168.3937.1338.

47. Wikipedia. Edward Anders, https://en.wikipedia.org/wiki/Edward_Anders (2024).

48. Sharov AA and Gordon R. Preface: Life as a cosmic phenomenon. In: Gordon R and Sharov AA (eds) *Habitability of the Universe Before Earth [in series: Astrobiology: Exploring Life on Earth and Beyond, eds Pabulo Henrique Rampelotto, Joseph Seckbach & Richard Gordon]*. Amsterdam: Elsevier B.V., 2018, pp. xxvii-xlii.

49. Archival Photographic File. Anders, Edward [apf1-00166], https://photoarchive.lib.uchicago.edu/db.xqy?one=apf1-00166.xml (1965).

50. Hayatsu R, Studier MH, Oda A, Fuse K and Anders E. Origin of organic matter in early solar system-II. Nitrogen compounds. *Geochim Cosmochim Acta* 1968; 32: 175–190.

51. Yoshino D, Hayatsu R and Anders E. Origin of organic matter in early solar system-III. Amino acids: catalytic synthesis. *Geochim Cosmochim Acta* 1971; 35: 927–938.

52. Hayatsu R, Studier MH and Anders E. Origin of organic matter in early solar system-IV. Amino acids: confirmation of catalytic synthesis by mass spectrometry. *Geochim Cosmochim Acta* 1971; 35: 939–951.

53. Studier MH, Anders E and Hayatsu R. Origin of organic matter in early solar system-V. Further studies of meteoritic hydrocarbons and a discussion of their origin. *Geochim Cosmochim Acta* 1972; 36: 189–215.

54. Hayatsu R, Studier MH, Matsuoka S and Anders E. Origin of organic matter in early solar system-VI. Catalytic synthesis of nitriles, nitrogen bases and prophyrin-like pigments. *Geochim Cosmochim Acta* 1972; 36: 555–571.

55. Hayatsu R, Matsuoka S, Scott RG, Studier MH and Anders E. Origin of organic matter in the early solar-system-VII. The organic polymer in carbonaceous chondrites. *Geochim Cosmochim Acta* 1977; 41: 1325–1339.

56. Wikipedia. Animal glue, https://en.wikipedia.org/wiki/Animal_glue (2024).

57. Mautner MN, Leonard RL and Deamer DW. Meteorite organics in planetary environments: hydrothermal release, surface activity, and microbial utilization. *Planet Space Sci* 1995; 43: 139–147. 1995/01/01.

58. Mautner MN. Biological potential of extraterrestrial materials .1. Nutrients in carbonaceous meteorites, and effects on biological growth. *Planet Space Sci* 1997; 45: 653–664. DOI: 10.1016/s0032-0633(97)00017-2.

59. Mautner MN, Conner AJ, Killham K and Deamer DW. Biological potential of extraterrestrial materials. 2. Microbial and plant responses to nutrients in the Murchison carbonaceous meteorite. *Icarus* 1997; 129: 245–253.

60. Gordon R and Sharov AA. Habitability of the Universe Before Earth [HUBE, Volume 1 in the series: Astrobiology: Exploring Life on Earth and Beyond, series editors: Pabulo Henrique Rampelotto, Joseph Seckbach & Richard Gordon]. In: Rampelotto PH, Seckbach J and Gordon R, (eds.). *Astrobiology: Exploring Life on Earth and Beyond.* Amsterdam: Elsevier B.V., 2018.

61. Wikipedia. Mechanical calculator, https://en.wikipedia.org/wiki/Mechanical_calculator (2024).

62. Barghoorn ES and Tyler SA. Microorganisms from Gunflint Chert: These structurally preserved precambrian fossils from Ontario are most ancient organisms known. *Science* 1965; 147: 563–575. DOI: 10.1126/science.147.3658.563.

63. Silverman SN, Kopf S, Gordon R, Bebout BM and Som SM. Measuring N_2 pressure using cyanobacteria [Abstract 4013]. In: Jang-Condell H (ed) *Habitable Worlds 2017: A System Science Workshop, November 13–17, 2017, University of Wyoming, Laramie, Wyoming.* Lunar and Planetary Institute, Universities Space Research Association, 2017, https://www.hou.usra.edu/meetings/habitableworlds2017/pdf/4013.pdf.

64. Silverman SN, Kopf S, Gordon R, Bebout B and Som S. Measuring ancient N_2 pressure using fossilized cyanobacteria In: Desch SJ (ed) *AbSciCon2017, Mesa, Arizona, April 24–28, 2017.* 2017, http://www.hou.usra.edu/meetings/abscicon2017/pdf/3242.pdf.

65. Silverman SN, Som S, Gordon R and Bebout B. Measuring ancient N_2 pressure using fossilized cyanobacteria *Rocky Mountain GeoBiology Symposium 2017, April 8, 2017*. Golden, CO, USA: Colorado School of Mines, 2017, https://rmgs2017.wordpress.com.

66. Silverman SN, Som S, Gordon R and Bebout B. Measuring ancient air pressure using fossilized cyanobacteria *Abstracts, American Geophysical Union Fall Meeting, San Francisco, 12–16 December 2016*. American Geophysical Union, 2016.

67. Walker DW. Obituary — James C. G. Wessel Walker. https://record.umich.edu/articles/obituary-james-c-g-wessel-walker/, 2023.

68. Ortiz MRD. Arecibo radio telescope. Arecibo Observatory, Puerto Rico, https://commons.wikimedia.org/wiki/File:Arecibo_radio_telescope_SJU_06_2019_6144.jpg (2019).

69. Gordon R and McNichol J. Recurrent dreams of life in meteorites. In: Seckbach J (ed) *Genesis — In the Beginning: Precursors of Life, Chemical Models and Early Biological Evolution*. Dordrecht, Netherlands: Springer, 2012, pp. 549–590.

70. Pillinger CT and Pillinger JM. A brief history of exobiology or there's nothing new in science. *Meteorit Planet Sci* 1997; 32: 443–445. Editorial Material. DOI: 10.1111/j.1945-5100.1997.tb01289.x.

71. McNichol J and Gordon R. Are we from outer space? A critical review of the panspermia hypothesis. In: Seckbach J (ed) *Genesis — In the Beginning: Precursors of Life, Chemical Models and Early Biological Evolution*. Dordrecht: Springer, 2012, pp. 591–620.

72. McNichol J. Jesse McNichol, https://www.stfx.ca/faculty-staff/jesse-mcnichol (2024).

73. Gordon R. *Origin of Life via Archaea: Shaped Droplets to Archaea First, With a Compendium of Archaea Micrographs [OOLA, Volume in the series Astrobiology Perspectives on Life of the Universe, Eds. Richard Gordon & Joseph Seckbach]*. Beverly, Massachusetts, USA: Wiley-Scrivener, 2024.

74. Burton AS and Berger EL. Insights into abiotically-generated amino acid enantiomeric excesses found in meteorites. *Life-Basel* 2018; 8. Review. DOI: 10.3390/life8020014.

75. Pappas S. 'Pristine' meteorite contaminated with table salt upon crash landing on Earth, https://www.livescience.com/pristine-meteorite-contaminated-with-table-salt-upon-crash-landing-on-earth (2023).

76. Wikipedia. 1997 Red River flood, https://en.wikipedia.org/wiki/1997_Red_River_flood (2024).

77. Wikipedia. Comet Hale–Bopp, https://en.wikipedia.org/wiki/Comet_Hale%E2%80%93Bopp (2024).

78. Winnipeg Free Press. *A Red Sea Rising: The Flood of the Century.* Winnipeg Free Press, 1997.

79. Gordon R. On stochastic growth and form. *Proceedings of the National Academy of Sciences USA* 1966; 56: 1497–1504.

80. Wikipedia. Aaron Novick, https://en.wikipedia.org/wiki/Aaron_Novick (2024).

81. Wikipedia. Ho Chi Minh, https://en.wikipedia.org/wiki/Ho_Chi_Minh (2024).

82. Stahl F. Aaron Novick (1919–2000), http://molbio.uoregon.edu/novick-history/ (2000).

83. Wikipedia. Leo Szilard, https://en.wikipedia.org/wiki/Leo_Szilard (2024).

84. Novick A and Szilard L. Experiments with the chemostat on spontaneous mutations of bacteria. *Proc Natl Acad Sci U S A* 1950; 36: 708–719. DOI: 10.1073/pnas.36.12.708.

85. Wikipedia. Directed evolution, https://en.wikipedia.org/wiki/Directed_evolution (2024).

86. Gordon R. Computer controlled evolution of diatoms: Design for a compustat. *Nova Hedwigia* 1996; 112: 215–219.

87. Pappas JL. *Mathematical Macroevolution in Diatom Research [MMDR, Volume in the series: Diatoms: Biology & Applications, series editors: Richard Gordon & Joseph Seckbach].* Beverly, MA, USA: Wiley-Scrivener, 2023.

88. Alivisatos P. Nanothermodynamics: A personal perspective by Terrell Hill. *Nano Lett* 2001; 1: 109. DOI: 10.1021/nl010012y.

89. Hersey J and Chappell W. *Hiroshima.* A. A. Knopf, 1946.

90. Lanoutte W and B. S. *Genius in the Shadows: A Biography of Leo Szilard, the Man Behind the Bomb.* New York: Charles Scribner's Sons, 1992.

91. Wikipedia. Peace Action. https://en.wikipedia.org/wiki/Peace_Action, 2024.

92. Gordon R and Chen Y. From statistical mechanics to molecular biology: A Festschrift for Terrell L. Hill. *Cell Biophys* 1987; 11: 0-iii.

93. Cheer A, Mogilner A and Stuchebrukhov A. In memoriam, Joel E. Keizer, Professor of Biological Sciences, UC Davis, 1942–1999, https://senate.universityofcalifornia.edu/_files/inmemoriam/html/joelekeizer.html (1999).

94. Meschel SV and Tarjan P. Transplanted Lives: The adventures of young Jewish immigrants from post-Fascist and Communist Hungary to the Free World following the 1956 Uprising https://www.amazon.com/Transplanted-Lives-adventures-immigrants-post-Fascist/dp/1537066765 (2016).

95. Maruyama M. The second cybernetics: Deviation-amplifying mutual causal processes. *Amer Sci* 1963; 51: 164–179.

96. Wikipedia. Magoroh Maruyama, https://en.wikipedia.org/wiki/Magoroh_Maruyama (2024).

97. Hill TL. *Thermodynamics of Small Systems, Part 1*. New York: W.A. Benjamin, 1963.

98. Hill TL. *Thermodynamics of Small Systems, Part 2*. New York: W.A. Benjamin, 1964.

99. Bedeaux D, Kjelstrup S and Schnell SK. *Nanothermodynamics*. World Scientific Press, 2023, p. 416.

100. Wikipedia. D'Arcy Wentworth Thompson, https://en.wikipedia.org/wiki/D'Arcy_Wentworth_Thompson (2024).

101. Thompson DAW. *On Growth and Form*. London, UK: Courier Corporation, 1942.

102. Gordon R. Adsorption isotherms of lattice gases by computer simulation. *J Chem Physics* 1968; 48: 1408–1409.

103. Gordon R. Steady-state properties of Ising lattice membranes. *J Chem Phys* 1968; 49: 570–580. DOI: 10.1063/1.1670111.

104. Braverman MH. Studies in hydroid differentiation II. Colony growth and the initiation of sexuality. *J Embryol Exp Morph* 1963; 11: 239–253.

105. Braverman MH and Schrandt RG. Computor generation of hypothetical hydroid colonies. *Am Zool* 1964; 4: 380–380.

106. Braverman MH and Schrandt RG. Colony development of a polymorphic hydroid as a problem in pattern formation. *Symp Zool Soc London* 1966; 16: 169–198.

107. Dignity Memorial. Magoroh Maruyama APRIL 2, 1929 – MARCH 16, 2018, https://www.dignitymemorial.com/obituaries/san-diego-ca/magoroh-maruyama-7793686 (2018). https://www.magiminiland.org/Days/

108. Baran RJ. Bonsai Book of Days: September. 2024.

109. Wikipedia. IBM hammer printers, https://en.wikipedia.org/wiki/IBM_hammer_printers (2024).

110. Wikipedia. Pauli effect, https://en.wikipedia.org/wiki/Pauli_effect (2023).

111. Church GM and Regis E. *Regenesis: How Synthetic Biology Will Reinvent Nature and Ourselves*. Basic Books, 2014.

112. Gordon R. *On Stochastic Growth and Form and Steady State Properties of Ising Lattice Membranes [Chemical Physics Ph.D. Thesis, Supervisor: Terrell L. Hill]*. Eugene, Oregon: Department of Chemistry, University of Oregon, 1967.

113. Hill TL and Kedem O. Studies in irreversible thermodynamics. III. Models for steady state and active transport across membranes. *J Theor Biol* 1966; 10: 399–441. DOI: 10.1016/0022-5193(66)90136-6.

114. Field RJ and Noyes RM. Oscillations in chemical systems .V. Quantitative explanation of band migration in the Belousov-Zhabotinskii reaction. *Journal of the American Chemical Society* 1974; 96: 2001–2006. DOI: 10.1021/ja00814a003.

115. Wikipedia. Boris Belousov (chemist), https://en.wikipedia.org/wiki/Boris_Belousov_(chemist) (2024).

116. Belousov BP. Периодически действующая реакция и ее механизм" [Periodically acting reaction and its mechanism]. *Сборник рефератов по радиационной медицине [Collection of abstracts on radiation medicine]* 1959; 147: 145–147.

117. Hill T. Studies in irreversible thermodynamics, VII. *Proc Natl Acad Sci U S A* 1966; 56: 840.

118. Ising E. Beitrag zur Theorie des Ferromagnetismus [Report on the theory of ferromagnetism]. *Z Phys* 1925; 31: 253–258. DOI: 10.1007/bf02980577.

119. Wikipedia. Ernst Ising, https://en.wikipedia.org/wiki/Ernst_Ising (2024).

120. Hill TL. Electric fields and cooperativity of biological membranes. *Proc Natl Acad Sci U S A* 1967; 58: 111-&. DOI: 10.1073/pnas.58.1.111.

121. Gordon R. Monte Carlo methods for cooperative Ising models. In: Karreman G (ed) *Cooperative Phenomena in Biology*. New York: Pergamon Press, 1980, pp. 189–241 + errata.

122. Teorell T. Electrokinetic consideration of mechanoelectrical transduction. *AnnNY AcadSci* 1966; 137: 950–966. DOI: 10.1111/j.1749-6632.1966.tb50209.x.

123. Jaynes ET, Rosenkrantz RD and Rosenkrantz RD. *E.T. Jaynes: Papers on Probability, Statistics, and Statistical Physics*. Springer Netherlands, 1989.

124. Wikipedia. Ora Kedem, https://en.wikipedia.org/wiki/Ora_Kedem (2023).

125. Gordon R. Steps in performing a 3-dimensional reconstruction of single asymmetric particles from a tilt series of electron micrographs. *Workshop on Information Treatment in Electron Microscopy, Basel* 1972.

126. Wikipedia. File:Ora Kedem.jpg, https://commons.wikimedia.org/wiki/File:Ora_Kedem.jpg (2023).

127. Breder Jr. CM. Fish schools as operational structures. *Fish Bull* 1976; 74: 471–502.

128. Clark E. Tribute to Dr. Charles M. Breder, Jr., http://isurus.mote.org/~kleber/tribute.htm (1991).

129. Calovi DS, Lopez U, Ngo S, Sire C, Chaté H and Theraulaz G. Swarming, schooling, milling: phase diagram of a data-driven fish school model. *New J Phys* 2014; 16: #015026. DOI: 10.1088/1367-2630/16/1/015026.

130. Newman JP and Sayama H. Effect of sensory blind zones on milling behavior in a dynamic self-propelled particle model. *Physical Review E* 2008; 78: #011913. DOI: 10.1103/PhysRevE.78.011913.

131. Lukeman R, Li YX and Edelstein-Keshet L. A conceptual model for milling formations in biological aggregates. *Bull Math Biol* 2009; 71: 352–382. DOI: 10.1007/s11538-008-9365-7.

132. Schneirla TC. A unique case of circular milling in ants, considered in relation to trail following and the general problem of orientation. *Amer Mus Novitates* 1944: 1–26.

133. Reynolds CW. Flocks, herds and schools: A distributed behavioral model. *SIGGRAPH Comput Graph* 1987; 21: 25–34.

134. Sivaram A and Venkatasubramanian V. Arbitrage equilibrium, invariance, and the emergence of spontaneous order in the dynamics of bird-like agents. *Entropy* 2023; 25: 15. DOI: 10.3390/e25071043.

135. Hermellin E and Michel F. Complex flocking dynamics without global stimulus. *14th European Conference on Artificial Life (ECAL)*. Natl Inst Appl Sci Lyon, Lyon, France: Mit Press, 2017, pp. 513–520.

136. Gordon R, Alicia. B. MM, Ather SH, Chen K, Konovalov AB, Lei P, Mazuritskiy M, Portygys T, Raj Vansh Singh S, Tot T, Ustaoglu Z, Valderrey ME, Vlasov VV and Wang C-X. *The End of Breast Cancer: Steerable Ray-by-Ray X-Ray Computed Tomography, [in preparation]*. Beverly, Massachusetts, USA: Wiley-Scrivener,, 2025.

137. Gordon R. Blinking Christmas tree lightbulbs in series. *Center for Theoretical Biology Quarterly Bulletin (Buffalo)* 1969; 2: 49–55.

138. Walmart. 2 Style D Flashers Christmas Tree Lights Replacement Bulbs 2.5v 0.425w Mini 0.17 a, https://www.walmart.com/ip/2-Style-D-Flashers-Christmas-Tree-Lights-Replacement-Bulbs-2-5v-0-425w-Mini-0-17-a/1006784475 (2024).

139. Stein WD. James Frederic Danielli, 1911–1984, Elected F.R.S. 1957. *Biogr Memoirs Fellows Roy Soc* 1986; 32: 115–135.

140. Rosen R. James F. Danielli: 1911–1984. *Journal of Social and Biological Structures* 1985; 8: 1–11.

141. Wikipedia. James Danielli, https://en.wikipedia.org/wiki/James_Danielli (2024).

142. Jeon KW, Lorch IJ and Danielli JF. Reassembly of living cells from dissociated components. *Science* 1970; 167: 1626–1627.

143. Gordon R, Bender R and Herman GT. What is ART? ART (Algebraic Reconstruction Techniques). *Center for Theoretical Biology Quarterly Bulletin* 1971; 4: (5 page insert).

144. King's College London. Professor James Danielli (1911–1984), https://www.kcl.ac.uk/people/james-danielli (1984).

145. Pincock S. Theodore Puck. *The Lancet* 2005; 366: 2000.

146. Lillie RS. The recovery of transmissivity in passive iron wires as a model of recovery processes in irritable living systems. Part I. *J Gen Physiol* 1920; 3: 107–128.

147. Lillie RS. The recovery of transmissivity in passive iron wires as a model of recovery processes in irritable living systems. Part II. *J Gen Physiol* 1920; 3: 129–143. DOI: 10.1085/jgp.3.2.129.

148. Lillie RS. Circuit transmission and interference of activation waves in living tissues and in passive iron. *Science* 1929; 69: 305–308. DOI: 10.1126/science.69.1785.305.

149. Gordon NK and Gordon R. *Embryogenesis Explained*. Singapore: World Scientific Publishing, 2017.

150. Cohn SA, Manoylov KM and Gordon R. Diatom Gliding Motility [DIGM, Volume in the series: Diatoms: Biology & Applications, series editors: Richard Gordon & Joseph Seckbach]. In: Cohn SA, Manoylov KM and Gordon R, (eds.). Beverly, MA, USA: Wiley-Scrivener, 2021.

151. Golomb SW. *Polyominoes*. Allen & Unwin, 1966.

152. Bitner JR and Reingold EM. Backtrack programming techniques. *Communications of the ACM* 1975; 18: 651–656.

153. Garner M. Mathematical Games: How rectangles, including squares, can be divided into squares of unequal size. *SciAm* 1958; 199: 136–142.

154. Mulcahy C and Richards D. Math Games of Martin Gardner Still Spur Innovation. https://www.scientifificamerican.com/article/math-games-of-martin-gardner-still-spur-innovation/, 2024.

155. Vukotić B and Gordon R. Habitable mini-Earths with black hole cores. In: Berea A (ed) *Technosignatures: Searching for and Communicating with Intelligence in Our Universe [TECH, Volume in the series Astrobiology Perspectives on Life of the Universe, Eds Richard Gordon & Joseph Seckbach]*. Beverly, Massachusetts, USA: Wiley-Scrivener, 2022, pp. 69–83.

156. Vukotić B, Seckbach J and Gordon R. Preface, Planet Formation and Panspermia: New Prospects for the Movement of Life through Space. *Planet Formation and Panspermia: New Prospects for the Movement of Life through Space [PNSP, Volume in the series Astrobiology Perspectives on Life of the Universe, Series Editors: Richard Gordon & Joseph Seckbach]*. Beverly, Massachusetts, USA: Wiley-Scrivener, 2021, pp. xi-xiv.

157. Vukotić B, Seckbach J and Gordon R. Planet Formation and Panspermia: New Prospects for the Movement of Life through Space [PNSP, Volume in the series Astrobiology Perspectives on Life of the Universe, Series

Editors: Richard Gordon & Joseph Seckbach]. Beverly, Massachusetts, USA: Wiley-Scrivener, 2021.

158. Vukotić B and Gordon R. Life in the Milky Way: The panspermia prospects [Chapter 4]. In: Vukotić B, Seckbach J and Gordon R (eds) *Planet Formation and Panspermia: New Prospects for the Movement of Life through Space [PNSP, Volume in the series Astrobiology Perspectives on Life of the Universe, Series Editors: Richard Gordon & Joseph Seckbach]*. Beverly, Massachusetts, USA: Wiley-Scrivener, 2021, pp. 41–52.

159. Jiřík J and Gordon R. Wet panspermia. In: Vukotić B, Seckbach J and Gordon R (eds) *Planet Formation and Panspermia: New Prospects for the Movement of Life through Space [PNSP, Volume in the series Astrobiology Perspectives on Life of the Universe, Series Editors: Richard Gordon & Joseph Seckbach]*. Beverly, Massachusetts, USA: Wiley-Scrivener, 2021, pp. 171–193.

160. Vukotić B and Gordon R. A survey of Solar System and galactic objects with pristine surfaces that record history and perhaps panspermia, with a plan for exploration [Chapter 14]. In: Vukotić B, Seckbach J and Gordon R (eds) *Planet Formation and Panspermia: New Prospects for the Movement of Life through Space [PNSP, Volume in the series Astrobiology Perspectives on Life of the Universe, Series Editors: Richard Gordon & Joseph Seckbach]*. Beverly, Massachusetts, USA: Wiley-Scrivener, 2021, pp. 269–308.

161. Gordon R. The panspermia publications of Sir Fred Hoyle [Chapter 15]. In: Vukotić B, Seckbach J and Gordon R (eds) *Planet Formation and Panspermia: New Prospects for the Movement of Life through Space [PNSP, Volume in the series Astrobiology Perspectives on Life of the Universe, Series Editors: Richard Gordon & Joseph Seckbach]*. Beverly, Massachusetts, USA: Wiley-Scrivener, 2021, pp. 309–326.

162. Gordon D. Gordon: You could learn to like crickets, https://www.westernstandard.news/opinion/gordon-you-could-learn-to-like-crickets/article_5736326a-2b02-11ed-9f75-bb3e848a8071.html (2022).

163. Pipes D and Gordon R. Will the Middle East Lose Its Importance?, http://www.danielpipes.org/blog/2012/07/will-the-middle-east-lose-its-importance (2012).

164. Gordon R. Yes to unilateral abandonement. Reader comment on item: [The Gaza Withdrawal:] A Democracy Killing Itself, http://www.danielpipes.org/comments/24665 (2005).

165. Gordon R. Boycott China. *Neepawa Banner & Press* 2020; 124: 5.

166. Gordon R. A poignant week in Vienna. *Der Standard/Transact — transnational activities in the cultural field in cooperation with museum in progress (Vienna)* 2000: 17.

167. Gordon R. Unilateral abandonment: Roadmap from the impasse between Israel and Palestine. *Jewish Post & News (Winnipeg)* 2004; 17: 5.

168. Gordon R. Afghans: no democracy for you, eh? *Uniter [University of Winnipeg]* 2006; 61: 7.

169. Gordon R. Accident gives Gordon a new view of university. *University of Manitoba Bulletin* 2002; 36: 5.

170. Spence S. Protest calls for peace in Afghanistan, https://uniter.ca/pdf/uniter-2006-11-02.pdf (2006).

171. Gordon R. Hoyle's tornado origin of artificial life, a computer programming challenge. In: Seckbach J and Gordon R (eds) *Divine Action and Natural Selection: Science, Faith and Evolution.* Singapore: World Scientific, 2008, pp. 354–367.

172. Hurn M. File:Institute of Astronomy, statue of Sir Fred Hoyle — geograph.org.uk — 372582.jpg, https://commons.wikimedia.org/wiki/File:Institute_of_Astronomy,_statue_of_Sir_Fred_Hoyle_-_geograph.org.uk_-_372582.jpg (2007).

173. Sharov AA and Gordon R. Life before Earth. In: Gordon R and Sharov AA (eds) *Habitability of the Universe Before Earth [in series: Astrobiology: Exploring Life on Earth and Beyond, eds Pabulo Henrique Rampelotto, Joseph Seckbach & Richard Gordon].* Amsterdam: Elsevier B.V., 2018, pp. 265–296.

174. Mikhailovsky G and Gordon R. The "Boring Billion" (1.8–0.8 Billion Years Ago) [BORB, Volume in the series Astrobiology Perspectives on Life of the Universe, Eds. Richard Gordon & Joseph Seckbach, in preparation]. Beverly, Massachusetts, USA: Wiley-Scrivener, 2025.

175. Dvorak R, Loibnegger B and Maindl TI. Possible origin of Theia, the Moon-forming impactor with Earth. *Astro Nachr* 2017; 338: 366–374. DOI: 10.1002/asna.201613209.

176. Williams DM and Zugger ME. Forming massive terrestrial satellites through binary-exchange capture. *Planet Sci J* 2024; 5: #208.

177. Gordon R and Mikhailovsky G. There were plenty of day/night cycles that could have accelerated an origin of life on Earth, without requiring panspermia [Chapter 11]. In: Vukotić B, Seckbach J and Gordon R (eds) *Planet Formation and Panspermia: New Prospects for the Movement of Life through Space [PNSP, Volume in the series Astrobiology Perspectives on Life of the Universe, Series Editors: Richard Gordon & Joseph Seckbach]*. Beverly, Massachusetts, USA: Wiley-Scrivener, 2021, pp. 195–206.

178. Gordon R and Mikhailovsky G. Appendix to Preface. In: Smoukov S, Seckbach J and Gordon R (eds) *Conflicting Models for the Origin of Life [COLF, Volume in the series: Astrobiology Perspectives on Life of the Universe, Series editors: Richard Gordon & Joseph Seckbach, Wiley-Scrivener]*. Beverly, Massachusetts, USA: Wiley-Scrivener, 2023, pp. xxiv-xxv.

179. Wikipedia. Polymerase chain reaction, https://en.wikipedia.org/wiki/Polymerase_chain_reaction (2024).

180. Smith J and Modrich P. Removal of polymerase-produced mutant sequences from PCR products. *Proc Natl Acad Sci U S A* 1997; 94: 6847–6850. DOI: 10.1073/pnas.94.13.6847.

181. Gordon R and Gordon NK. How to make a transmembrane domain at the origin of life: A possible origin of proteins [Chapter 7]. In: Smoukov SK, Seckbach J and Gordon R (eds) *Conflicting Models for the Origin of Life [COLF, Volume in the series Astrobiology Perspectives on Life of the Universe, Eds Richard Gordon & Joseph Seckbach]*. Beverly, Massachusetts, USA: Wiley-Scrivener, 2023, pp. 131–174.

182. Tirunelveli G, Gordon R and Pistorius S. Comparison of square-pixel and hexagonal-pixel resolution in image processing. In: Kinsner W, Sebak A and Ferens K (eds) *IEEE CCEC 2002: Canadian Conference on Electrical and Computer Engineering, Vols 1–3, Conference Proceedings*. 2002, pp. 867–872.

183. Gordon R. *The Hierarchical Genome and Differentiation Waves: Novel Unification of Development, Genetics and Evolution*. Singapore & London: World Scientific & Imperial College Press, 1999.

184. Gordon R and Mikhailovsky G. Continuing differentiation and the multiple origins of complex multicellularity. In: Mikhailovsky G and Gordon R (eds) *The "Boring Billion" (18–08 Billion Years Ago) [BORB, Volume in the series Astrobiology Perspectives on Life of the Universe, Eds Richard Gordon & Joseph Seckbach]*. Beverly, Massachusetts, USA: Wiley-Scrivener, 2025, pp. in preparation.

185. Open Access Pub. George Mikhailovsky, https://oap-lifescience.org/evolutionary-science/evolution/george-mikhailovsky (2024).

186. Gordon R. Polyribosome dynamics at steady state. *J Theor Biol* 1969; 22: 515–532.

187. Miller C. File:Multiple Ribosomes Translation Protein Synthesis. https://commons.wikimedia.org/wiki/File:Multiple_Ribosomes_Translation_Protein_Synthesis.png (2020).

188. Wikipedia. Luna Leopold, https://en.wikipedia.org/wiki/Luna_Leopold (2023).

189. Wikipedia. Kinematic wave, https://en.wikipedia.org/wiki/Kinematic_wave (2023).

190. Wikipedia. Hans Albert Einstein, https://en.wikipedia.org/wiki/Hans_Albert_Einstein (2024).

191. Gordon R, Carmichael JB and Isackson FJ. Saltation of plastic balls in a 'one-dimensional' flume. *Water Resources Res* 1972; 8: 444–459.

192. Chevalier C, Dorignac J, Ibrahim Y, Choquet A, David A, Ripoll J, Rivals E, Geniet F, Walliser N-O, Palmeri J, Parmeggiani A and Walter J-C. Physical modeling of ribosomes along messenger RNA: Estimating kinetic parameters from ribosome profiling experiments using a ballistic model. *PLoS Comput Biol* 2023; 19: #e1011522. DOI: 10.1371/journal.pcbi.1011522.

193. Bender R, Bellman SH and Gordon R. ART and the ribosome: A preliminary report on the three-dimensional structure of individual ribosomes determined by an Algebraic Reconstruction Technique. *J Theor Biol* 1970; 29: 483–488.

194. Cropo Funeral Chapel. Obituary of Frans Visscher, https://cropo.com/tribute/details/3628/Frans-Visscher/obituary.html (2014).

195. Wikipedia. Humberto Fernández-Morán, https://en.wikipedia.org/wiki/Humberto_Fern%C3%A1ndez-Mor%C3%A1n (2023).

196. Wikipedia. Albert Crewe, https://en.wikipedia.org/wiki/Albert_Crewe (2023).

197. Czarnecki M. New Life for Sale: Mass-produced life goes to market. *Maclean's* 1980: 41–49.

198. Wikipedia. Aaron Klug, https://en.wikipedia.org/wiki/Aaron_Klug (2024).

199. Wikipedia. Cyrus Levinthal, https://en.wikipedia.org/wiki/Cyrus_Levinthal (2023).

200. wikiHow Staff. How to Rake a Zen Garden, https://www.wikihow.com/Rake-a-Zen-Garden (2023).

201. Gordon R. Stop breast cancer now! Imagining imaging pathways towards search, destroy, cure and watchful waiting of premetastasis breast cancer. In: Tot T (ed) *Breast Cancer — A Lobar Disease*. London: Springer, 2011, pp. 167–203.

202. Gordon R and Herman GT. Reconstruction of pictures from their projections. *Communications of the ACM* 1971; 14: 759–768. DOI: 10.1145/362919.362925.

203. Gordon R, Rowe Jr JE and Bender R. ART: a possible replacement for x-ray crystallography at moderate resolution. In: Broda E, Locker A and Springer-Lederer H (eds) *Proceedings of the First European Biophysics Congress, Vol VI: Theoretical Molecular Biology, Biomechanics, Biomathematics, Environmental Biophysics, Techniques, Education*. Vienna: Verlag der Wiener Medizinischer Akademie, 1971, pp. 60, 441–445.

204. Chemistry Tree. Cyrus Levinthal, https://academictree.org/chemistry/peopleinfo.php?pid=5408 (1990).

205. Cobb M and Comfort N. What Rosalind Franklin truly contributed to the discovery of DNA's structure. *Nature* 2023; 616: 657–660. DOI: 10.1038/d41586-023-01313-5.

206. Bellman SH, Bender R, Gordon R and Rowe Jr JE. ART is science, being a defense of Algebraic Reconstruction Techniques for three-dimensional electron microscopy. *J Theor Biol* 1971; 32: 205–216.

207. Carroll L. The Hunting of the Snark, https://www.poetryfoundation.org/poems/43909/the-hunting-of-the-snark (1876).

208. Gordon R, Bender R and Herman GT. Algebraic Reconstruction Techniques (ART) for three-dimensional electron microscopy and x-ray photography. *J Theor Biol* 1970; 29: 471–481.

209. Gilbert P. Iterative methods for the three-dimensional reconstruction of an object from projections. *J Theor Biol* 1972; 36: 105–117.

210. Gordon R and Carron I. SAFIRE (Sinogram Affirmed Iterative Reconstruction), http://nuit-blanche.blogspot.com/2012/02/les-cameras-aleatoires-and-nuit.html?utm_source=feedburner&utm_medium=email&utm_campaign=Feed%3A+blogspot%2FvhVI+%28Nuit+Blanche%29 (2012).

211. Nakazato Y and Otaki JM. Socket array irregularities and wing membrane distortions at the eyespot foci of butterfly wings suggest mechanical signals for color pattern determination. *Insects* 2024; 15: #535.

212. Van Noorden R. Ribosome clinches the chemistry Nobel. *Nature* 2009. DOI: 10.1038/news.2009.981.

213. Gordon R. Reconstruction from projections: A survey of applications. *Proceedings of the Fourth International Congress for Stereology, NBS, Gaithersburg, Maryland, September 4–9, 1975 [National Bureau of Standards Special Publication 431]*. Gaithersburg, MD USA: National Bureau of Standards, 1976, pp. 201–202.

214. Gordon R and Bender R. The ART of sectioning without cutting. In: Weibel ER (ed) *Third International Congress for Stereology, Berne, 25–31 August 1971 Abstracts*. 1971, pp. 17.

215. Kaczmarz S. Angenäherte auflösung von systemen linearer Gleichungen. *Bulletin International de l' Académie Polonaise des Sciences et des Lettres* 1937; 35: 355–357.

216. Colquhoun GD and Gordon R. The use of control angles with MART (Multiplicative Algebraic Reconstruction Technique). *Technology in Cancer Research and Treatment* 2005; 4: 183–184.

217. Beyer WA and Ulam SM. Note on the visual hull of a set. *Journal of Combinatorial Theory* 1968; 4: 240–245.

218. Meisters GH and Ulam SM. On visual hulls of sets. *Proceedings of the National Academy of Sciences USA* 1967; 57: 1172–1174.

219. Gordon R and Colquhoun GD. *CancerZap!*: Battleship meets *Where's Waldo? BioPhotonics* 2012; 19: 8.

220. Colquhoun GD and Gordon R. Foxels for low dose pediatric CT. *Manitoba Institute of Child Health 2nd Annual Research Day Growing*

Up — Growing Better: A Celebration of Manitoba Child Health Research, November 16, 2006. 2006.

221. Colquhoun GD and Gordon R. A superresolution computed tomography algorithm for reverse cone beam 3D x-ray mammography [PowerPoint presentation]. In: Tot T (ed) *Workshop on Alternatives to Mammography, Copenhagen, September 29–30, 2005.* Alonsa, Manitoba, Canada: Silver Bog Research Inc., 2005.

222. Potter MJA, Colquhoun GD and Gordon R. Design of a 3D Microtumour Breast Scanner Using 7th-Generation CT (Computed Tomography) [PowerPoint presentation]. In: Gumel A, Guo C-H, Li C, *et al.* (eds) *Fourth Annual Meeting of the Prairie Network for Research in Mathematical Sciences (PNRMS), University of Manitoba, April 30-May 2, 2010.* Winnipeg: Prairie Network for Research in Mathematical Sciences, 2010.

223. Wikipedia. Joachim Frank, https://en.wikipedia.org/wiki/Joachim_Frank (2023).

224. Royal Swedish Academy of Sciences. The Nobel Prize in Chemistry, (2017).

225. Kuhn TS. *The Structure of Scientific Revolutions.* University of Chicago Press, 1996.

226. Wikipedia. Thomas Kuhn, https://en.wikipedia.org/wiki/Thomas_Kuhn (2024).

227. Goel NS, Campbell RD, Gordon R, Rosen R, Martinez H and Ycas M. Self-sorting of isotropic cells. *J Theor Biol* 1970; 28: 423–468.

228. Rosen R. *Life Itself: A Comprehensive Inquiry Into the Nature, Origin, and Fabrication of Life.* Columbia University Press, 1991.

229. Rosen R. *Optimality Principles in Biology.* Springer US, 1967.

230. Gordon R, Goel NS, Steinberg MS and Wiseman LL. A rheological mechanism sufficient to explain the kinetics of cell sorting. *J Theor Biol* 1972; 37: 43–73.

231. Gordon R, Goel NS, Steinberg MS and Wiseman LL. A rheological mechanism sufficient to explain the kinetics of cell sorting. In: Mostow GD (ed) *Mathematical Models for Cell Rearrangement.* New Haven: Yale University Press, 1975, pp. 196–230.

232. Goel NS, Campbell RD, Gordon R, Rosen R, Martinez H and Ycas M. Self-sorting of isotropic cells. In: Mostow GD (ed) *Mathematical Models for Cell Rearrangement*. New Haven: Yale University Press, 1975, pp. 100–144.

233. Steinberg MS. Reconstruction of tissues by dissociated cells. *Science* 1963; 141: 401–408. DOI: 10.1126/science.141.3579.401.

234. Phillips HM, Steinberg MS and Lipton BH. Embryonic tissues as elastoviscous liquids .2. Direct evidence for cell slippage in centrifuged aggregates. *Dev Biol* 1977; 59: 124–134. DOI: 10.1016/0012-1606(77)90247-0.

235. Phillips HM and Steinberg MS. Embryonic tissues as elastoviscous liquids .1. Rapid and slow shape changes in centrifuged cell aggregates. *J Cell Sci* 1978; 30: 1–20.

236. Wikipedia. Aron Moscona, https://en.wikipedia.org/wiki/Aron_Moscona (2023).

237. Moscona AA. The development in vitro of chimeric aggregates of dissociated embryonic chick and mouse cells. *Proc Natl Acad Sci USA* 1957; 43: 184–194.

238. Goldschneider I and Moscona AA. Tissue-specific cell-surface antigens in embryonic cells. *Journal of Cell Biology* 1972; 53: 435–449. DOI: 10.1083/jcb.53.2.435.

239. Steinberg MS. Differential adhesion in morphogenesis: a modern view. *Curr Opin Genet Dev* 2007; 17: 281–286. DOI: 10.1016/j.gde.2007.05.002.

240. Schötz EM, Burdine RD, Jülicher F, Steinberg MS, Heisenberg CP and Foty RA. Quantitative differences in tissue surface tension influence zebrafish germ layer positioning. *Hfsp J* 2008; 2: 42–56. DOI: 10.2976/1.2834817.

241. Oriola D, Marin-Riera M, Anlaş K, Gritti N, Sanaki-Matsumiya M, Aalderink G, Ebisuya M, Sharpe J and Trivedi V. Arrested coalescence of multicellular aggregates. *Soft Matter* 2022; 18: 3771–3780. 10.1039/D2SM00063F. DOI: 10.1039/D2SM00063F.

242. Belousov R, Savino S, Moghe P, Hiiragi T, Rondoni L and Erzberger A. Poissonian cellular Potts models reveal nonequilibrium kinetics of cell sorting. *Physical Review Letters* 2024; 132: 7. DOI: 10.1103/PhysRevLett.132.248401.

243. Dahl-Jensen S and Grapin-Botton A. The physics of organoids: A biophysical approach to understanding organogenesis. *Development* 2017; 144: 946–951. DOI: 10.1242/dev.143693.

244. Chalkley HW. Quantitative relation between the number of organized centers and tissue volume in regenerating masses of minced body sections of hydra. *Natl Cancer Inst J* 1945; 6: 191–195.

245. Gierer A, Bode H, Berking S, Schaller H, Trenkner E, Hansmann G, David CN and Flick K. Regeneration of hydra from reaggregated cells. *Nature-New Biology* 1972; 239: 98–101.

246. Galtsoff PS. Regeneration after dissociation (An experimental study on sponges) II Histogenesis of *Microciona prolifera*, Verf. *J Exp Zool* 1925; 42: 223–255. DOI: 10.1002/jez.1400420110.

247. Humphreys T, Humphreys S and Moscona AA. A procedure for obtaining completely dissociated sponge cells [abstract]. *Biol Bull* 1960; 119: 294.

248. Wikipedia. Malcolm Steinberg, https://en.wikipedia.org/wiki/Malcolm_Steinberg (2024).

249. Kelly M. Renowned Princeton biologist Malcolm Steinberg dies, https://www.princeton.edu/news/2012/02/22/renowned-princeton-biologist-malcolm-steinberg-dies (2012).

250. Harvey EN. Observations on living cells, made with the microscope-centrifuge. *J Exp Biol* 1931; 8: 267–274.

251. Phillips HM, Steinberg MS and Lipton BH. Elastoviscous morphogenetic behavior of centrifuged embryonic chick cell aggregates [abstract]. *Am Zool* 1972; 12: 699.

252. Gordon R. On Monte Carlo algebra. *J Appl Probab* 1970; 7: 373–387.

253. Gordon R. *Lattice Membrane with Nearest Neighbor Interactions [16mm movie]*. Eugene: University of Oregon, 1966.

254. Hammersley J. *Monte Carlo Methods*. Methuen, 1964.

255. Waffle B. Mini Beam – A Compact X-ray Microbeam Generator for Micro X-ray Analysis (Aug 2021), https://www.youtube.com/watch?v=Il6Cdq8l70Y (2021).

256. Zhou XH and Gordon R. Detection of early breast cancer: An overview and future prospects. *Crit Rev Biomed Eng* 1989; 17: 203–255.

257. Ather SH and Gordon R. Electronic steering of x-rays. *Journal of Artificial Intelligence and Robotics*, 1(2), https://joaiar.org/articles/AIR-1012.pdf. 2024.

258. Gordon R and Drum RW. A capillarity mechanism for diatom gliding locomotion. *Proc Natl Acad Sci U S A* 1970; 67: 338–344.

259. Kraberg A. *Bacillaria paxillifera* (O.F.Müller) T.Marsson, 1901, https://planktonnet.awi.de/index.php?contenttype=image_details&itemid=62887#content (2014).

260. Bonner JT. *Morphogenesis: An Essay on Development*. Atheneum, 1963.

261. Sabuncu AC, Gordon R, Richer E, Manoylov KM and Beskok A. The kinematics of explosively jerky diatom motility: A natural example of active nanofluidics [Chapter 2]. In: Cohn SA, Manoylov KM and Gordon R (eds) *Diatom Gliding Motility [Volume in the series: Diatoms: Biology & Applications, series editors: Richard Gordon & Joseph Seckbach]*. Beverly, MA, USA: Wiley-Scrivener, 2021, pp. 33–63.

262. Alicea B, Gordon R and Parent J. The psychophysical world of the motile diatom *Bacillaria paradoxa* [Chapter 9]. In: Pappas JL and Gordon R (eds) *The Mathematical Biology of Diatoms [DMTH, Volume in the series: Diatoms: Biology & Applications, series editors: Richard Gordon & Joseph Seckbach]*. Beverly, MA, USA: Wiley-Scrivener, 2023, pp. 229–263.

263. Ussing AP, Gordon R, Ector L, Buczkó K, Desnitskiy AG and VanLandingham SL. *The Colonial Diatom "Bacillaria paradoxa": Chaotic Gliding Motility, Lindenmeyer Model of Colonial Morphogenesis, and Bibliography, with Translation of O.F. Müller (1783), "About a peculiar being in the beach-water" [Diatom Monographs, vol. 5, Witkowski, Andrzej, editor]*. Ruggell: A.R.G. Ganter Verlag Kommanditgesellschaft, 2005, p. 139.

264. Alicea B, Gordon R, Harbich T, Singh U, Sing A and Varma V. Towards a digital diatom: Image processing and deep learning analysis of *Bacillaria paradoxa* dynamic morphology [Chapter 10]. In: Cohn SA, Manoylov KM and Gordon R (eds) *Diatom Gliding Motility [DIGM, Volume in the series: Diatoms: Biology & Applications, series editors: Richard Gordon & Joseph Seckbach]*. Beverly, MA, USA: Wiley-Scrivener, 2021, pp. 223–248.

265. Gordon R and Aguda BD. Diatom morphogenesis: Natural fractal fabrication of a complex microstructure. In: Harris G and Walker C

(eds) *Proceedings of the Annual International Conference of the IEEE Engineering in Medicine and Biology Society, Part 1/4: Cardiology and Imaging, 4–7 Nov 1988, New Orleans, LA, USA*. New York: Institute of Electrical and Electronics Engineers, 1988, pp. 273–274.

266. Gautam S, Kashyap M, Gupta S, Kumar V, Schoefs B, Gordon R, Joshi KB, Vinayak V and Jeffryes C. Metabolic engineering of TiO_2 nanoparticles in *Nitzschia palea* to form diatom nanotubes: an ingredient for solar cells to produce electricity and biofuel. *RSC Adv* 2016; 6: 97276–97284.

267. Vinayak V, Joshi KB, Gordon R and Schoefs B. Nanoengineering of diatom surfaces for emerging applications [Chapter 3]. In: Losic D (ed) *Diatom Nanotechnology: Progress and Emerging Applications*. The Royal Society of Chemistry, 2018, pp. 55–78.

268. Ghobara MM, Mazumder N, Vinayak V, Reissig L, Gebeshuber IC, Tiffany MA and Gordon R. On light and diatoms: A photonics and photobiology review [Chapter 7]. In: Seckbach J and Gordon R (eds) *Diatoms: Fundamentals & Applications [DIFA, Volume 1 in the series: Diatoms: Biology & Applications, series editors: Richard Gordon & Joseph Seckbach]*. Beverly, MA, USA: Wiley-Scrivener, 2019, pp. 129–190.

269. Gordon R, Katyal K and Raj Vansh Singh S. An attempted simulation of diatom motility at the molecular level: The domino effect hydration model with concerted diffusion, https://isdr.org/wp-content/uploads/2021/08/Online-IDS-Oral-program.pdf (2021).

270. Raj Vansh Singh S, Kashyap K and Gordon R. RAPHE: Simulation of the dynamics of diatom motility at the molecular level — The domino effect hydration model with concerted diffusion [Chapter 11]. In: Pappas JL (ed) *The Mathematical Biology of Diatoms[DMTH, Volume in the series: Diatoms: Biology& Applications, series editors: Richard Gordon & Joseph Seckbach]*. Beverly, Massachusetts, USA: Wiley-Scrivener, 2023, pp. 291–342.

271. Gordon R. The whimsical history of proposed motors for diatom motility [Chapter 14]. In: Cohn SA, Manoylov KM and Gordon R (eds) *Diatom Gliding Motility [DIGM, Volume in the series: Diatoms: Biology & Applications, series editors: Richard Gordon & Joseph Seckbach]*. Beverly, MA, USA: Wiley-Scrivener, 2021, pp. 335–420.

272. Cohn SA, Manoylov KM and Gordon R. Preface. In: Cohn SA, Manoylov KM and Gordon R (eds) *Diatom Gliding Motility [DIGM, Volume in*

the series: *Diatoms: Biology & Applications, series editors: Richard Gordon & Joseph Seckbach]*. Beverly, MA, USA: Wiley-Scrivener, 2021, pp. xxvii-xxxvi.

273. Parsons DF. Electron microscopy, electron diffraction, and element analysis of wet biological specimens. *Cell Biophys* 1988; 13: 159–171.

274. Gordon R, Drum RW and Brownstone R. Oscillatory behaviour of single cells of the colonial diatom *Bacillaria paradoxa*. In: Patrick R and Reimer CW (eds) *VII International Symposium on Living and Fossil Diatoms, August 22–28, 1982, Abstracts of Posters*. Philadelphia: Academy of Natural Sciences, 1982.

275. Gordon R. Partial synchronization of the colonial diatom *Bacillaria "paradoxa"*. *Research Ideas and Outcomes (RIO)* 2016; 2. DOI: 10.3897/rio.2.e7869.

276. Kapinga MRM and Gordon R. Cell motility rhythms in *Bacillaria paxillifer*. *Diatom Res* 1992; 7: 221–225.

277. Kapinga M. The motility of *Bacillaria paradoxa*. In: Roeder T (ed) *IX North American Diatom Symposium, Treehaven Field Station, Tomahawk, WI 54487, October 7–10, 1987*. Stevens Point: University of Wisconsin, 1987, abstract #18.

278. Kapinga M and Gordon R. Cell to cell communication in the gliding diatom *Bacillaria. Proceedings of the Microscopical Society of Canada, Fourteenth Annual Meeting, 17–19 June 1987, University of Manitoba, Winnipeg, Manitoba*. Toronto: Microscopical Society of Canada, 1987, p. 12.

279. Kapinga MRM. *Observations on the Growth and Motile Behavior of the Colonial Diatom Bacillaria paradoxa in Culture [M.Sc. Thesis, Supervisor: R. Gordon]*. Winnipeg: University of Manitoba, 1989.

280. Kapinga MRM and Gordon R. Cell attachment in the motile colonial diatom *Bacillaria paxillifer*. *Diatom Res* 1992; 7: 215–220.

281. Gordon R. The Glass Menagerie: Diatom Nanotechnology and Its Implications for Multi-Scale Manufacturing and Oil Production [PowerPoint presentation]. In: Raut JS and Venkataraghavan R (eds) *2nd International Conference on Multi-Scale Structures and Dynamics of Complex Systems: Processes & Forces for Creation of Designer*

Materials with Multi-Scale Structures, 4–5 September, 2008, Bangalore, India. Bangalore: Hindustan Unilever Ltd., 2008.

282. Singh U, Alicea B, Gordon R and Harbich T. Unravelling the behavioral psychology of *Bacillaria* cells in gliding chains. In: Congestri R, Seckbach J and Gordon R (eds) *Chain Diatoms [DCHN, Volume in the series: Diatoms: Biology & Applications, series editors: Richard Gordon & Joseph Seckbach]*. Beverly, MA, USA: Wiley-Scrivener, 2020, In preparation.

283. Harbich T. Modeling the synchronization of the movement of *Bacillaria paxillifer* by a Kuramoto Model with time delay. In: Pappas JL (ed) *The Mathematical Biology of Diatoms [DMTH, Volume in the series: Diatoms: Biology & Applications, series editors: Richard Gordon & Joseph Seckbach]*. Beverly, Massachusetts, USA: Wiley-Scrivener, 2023, pp. 193–228.

284. Gordon R, Losic D, Tiffany MA, Nagy SS and Sterrenburg FAS. The Glass Menagerie: Diatoms for novel applications in nanotechnology. *Trends Biotechnol* 2009; 27: 116–127. DOI: https://doi.org/10.1016/j.tibtech.2008.11.003.

285. Cohn SA, Spurck TP and Pickett-Heaps JD. High energy irradiation at the leading tip of moving diatoms causes a rapid change of cell direction. *Diatom Res* 1999; 14: 193–206.

286. Gordon R. *Diatom Nanotechnology, Nanofluidics and Photonics [PowerPoint presentation, November 30, 2011]*. Québec, PC, Canada: Centre for Optics, Photonics and Laser, University of Laval, 2011.

287. Gordon R and Brodland GW. On square holes in pennate diatoms. *Diatom Res* 1990; 5: 409–413.

288. Gordon R and Drum RW. The chemical basis of diatom morphogenesis. *International Review of Cytology* 1994; 150: 243–372, 421–422. DOI: 10.1016/S0074-7696(08)61544-2.

289. Parkinson J, Brechet Y and Gordon R. Centric diatom morphogenesis: A model based on a DLA algorithm investigating the potential role of microtubules. *Biochimica et Biophysica Acta (BBA) — Molecular Cell Research* 1999; 1452: 89–102. DOI: https://doi.org/10.1016/S0167-4889(99)00116-0.

290. Round FE, Crawford RM and Mann DG. *The Diatoms, Biology & Morphology of the Genera*. Cambridge: Cambridge University Press, 1990.

291. Annenkov VV, Gordon R, Zelinskiy SN and Danilovtseva EN. Endocytosis is the probable mechanism for silicon capture by diatom algae — Is endocytosis a key stage in building of siliceous frustules? *J Phycol* 2020; 50: DOI: 10.1111/jpy.13062.

292. Annenkov VV, Danilovtseva EN and Gordon R. Steps of silicic acid transformation to silica frustules: Main hypothesis and discoveries [Chapter 13]. In: Annenkov V, Seckbach J and Gordon R (eds) *Diatom Morphogenesis [DIMO, Volume in the series: Diatoms: Biology & Applications, series editors: Richard Gordon & Joseph Seckbach]*. Beverly, MA, USA: Wiley-Scrivener, 2021, pp. 301–347.

293. Annenkov VV, Seckbach J and Gordon R. Diatom Morphogenesis [DIMO, Volume in the series: Diatoms: Biology & Applications, series editors: Richard Gordon & Joseph Seckbach]. In: Annenkov V, Seckbach J and Gordon R, (eds.). Beverly, MA, USA: Wiley-Scrivener, 2021.

294. Mazumder N and Gordon R. Diatom Microscopy [DIMI, Volume in the series: Diatoms: Biology & Applications, series editors: Richard Gordon & Joseph Seckbach]. In: Gordon R and Seckbach J, (eds.). *Diatoms: Biology & Applications*. Beverly, MA, USA: Wiley-Scrivener, 2022.

295. Pappas JL. The Mathematical Biology of Diatoms [DMTH, Volume in the series: Diatoms: Biology & Applications, series editors: Richard Gordon & Joseph Seckbach]. Beverly, MA, USA: Wiley-Scrivener, 2023.

296. Seckbach J and Gordon R. Diatoms: Fundamentals & Applications [DIFA, Volume 1 in the series: Diatoms: Biology & Applications, series editors: Richard Gordon & Joseph Seckbach]. In: Gordon R and Seckbach J, (eds.). *Diatoms: Biology & Applications*. Beverly, MA, USA: Wiley-Scrivener, 2019.

297. Vinayak V and Gordon R. Diatom Cultivation for Biofuel, Food and High Value Products [DCUL, Volume in the series: Diatoms: Biology & Applications, series editors: Richard Gordon & Joseph Seckbach, in press]. In: Gordon R and Seckbach J, (eds.). Beverly, MA, USA: Wiley-Scrivener, 2024.

298. Maidana NI, Licursi M and Morales EA. Diatom Ecology: From Molecules to Metacommunities [DIEC, Volume in the series: Diatoms: Biology & Applications, series editors: Richard Gordon & Joseph Seckbach]. In: Gordon R and Seckbach J, (eds.). *Diatoms: Biology & Applications*. Beverly, MA, USA: Wiley-Scrivener, 2024.

299. Pappas JL, Tiffany MA and Gordon R. The uncanny symmetry of some diatoms and not of others: A multi-scale morphological characteristic and a puzzle for morphogenesis [Chapter 2]. In: Annenkov V, Seckbach J and Gordon R (eds) *Diatom Morphogenesis [DIMO, Volume in the series: Diatoms: Biology & Applications, series editors: Richard Gordon & Joseph Seckbach]*. Beverly, MA, USA: Wiley-Scrivener, 2021, pp. 19–67.

300. Pappas JL and Gordon R. Are mantle profiles of circular centric diatoms a measure of buckling forces during valve morphogenesis? [Chapter 9]. In: Annenkov V, Seckbach J and Gordon R (eds) *Diatom Morphogenesis [DIMO, Volume in the series: Diatoms: Biology & Applications, series editors: Richard Gordon & Joseph Seckbach]*. Beverly, MA, USA: Wiley-Scrivener, 2021, pp. 231–250.

301. University of Waterloo. Wayne Brodland, PEng, https://uwaterloo.ca/civil-environmental-engineering/profile/brodland (2024).

302. Gordon R and Walsby AE. Development of the heterocyst spacing problem in *Anabaena. Serbelloni Symposium on Theoretical Biology, June, 1970.* 1970.

303. Bdcarl. *Anabaena circinalis* filament, https://commons.wikimedia.org/w/index.php?curid=19074360 (2023).

304. Bonner JT and Hoffman ME. Evidence for a substance responsible for the spacing pattern of aggregation and fruiting in the cellular slime molds. *Journal of Embryology and Experimental Morphology* 1963; 11: 571–589.

305. Baker RW and Herman GT. Simulation of organisms using a developmental model. 1. Basic description. *Int J Biomed Comput* 1972; 3: 201–215.

306. Baker RW and Herman GT. Simulation of organisms using a developmental model. Part 2: The heterocyst formation problem in blue-green algae. *Int J Biomed Comput* 1972; 3: 251–267.

307. Walsby AE. A square bacterium. *Nature* 1980; 283: 69–71.

308. Walsby AE. Archaea with square cells. *Trends Microbiol* 2005; 13: 193–195. DOI: 10.1016/j.tim.2005.03.002.

309. Walsby AE. Mucilage secretion and the movements of blue-green algae. *Protoplasma* 1968; 65: 223–238.

310. Walsby AE. Gordon Elliott Fogg CBE 26 April 1919 — 30 January 2005. *Biogr Mems Fell R Soc* 2006; 52: 97–116.

311. Wikipedia. *Haloquadratum walsbyi*, https://en.wikipedia.org/wiki/Haloquadratum_walsbyi (2024).

312. Gordon R and Bender R. New three-dimensional Algebraic Reconstruction Techniques (ART). *Proceedings, Twenty-ninth Annual Meeting, Electron Microscopy Society of America, Boston, Massachusetts, August 9, 10, 11, 12 and 13, 1971*. Baton Rouge: Claitor's Publishing Division, 1971, pp. 82–83.

313. Honda H. Description of form of trees by parameters of tree-like body: Effects of branching angle and branch length on shape of tree-like body. *J Theor Biol* 1971; 31: 331–338. DOI: 10.1016/0022-5193(71)90191-3.

314. Leopold LB. Trees and streams: Efficiency of branching patterns. *J Theor Biol* 1971; 31: 339–354. DOI: 10.1016/0022-5193(71)90192-5.

315. Delagrange S, Jauvin C and Rochon P. PypeTree: A tool for reconstructing tree perennial tissues from point clouds. *Sensors (Basel, Switzerland)* 2014; 14: 4271–4289. DOI: 10.3390/s140304271.

316. Kilmer J. Trees, https://www.poetryfoundation.org/poetrymagazine/poems/12744/trees (1913).

317. Carmichael J. Dr. Jack Carmichael, Michigan State University alumnus and noted chemist and expert in environmental pollution studies and industrial development discusses his life and career in a wide-ranging oral history interview [MP3 audio], https://d.lib.msu.edu/vvl/3969 (2015).

318. Fang Y, Lu Y, Roskilly AP and Yu X. A review of compressed air energy systems in vehicle transport. *Energy Strategy Reviews* 2021; 33: 100583. DOI: https://doi.org/10.1016/j.esr.2020.100583.

319. Wikipedia. Paul Flory, https://en.wikipedia.org/wiki/Paul_Flory (2024).

320. Wikipedia. Ralph Bagnold, https://en.wikipedia.org/wiki/Ralph_Bagnold (2023).

321. Bagnold R. *The Physics of Blown Sand and Desert Dunes*. Springer Netherlands, 2012.

322. Moreland EL, Arvidson RE, Morris RV, Condus T, Hughes MN, Weitz CM and VanBommel SJ. Orbital and in situ investigation of the Bagnold Dunes and Sands of Forvie, Gale Crater, Mars. *Journal of Geophysical Research: Planets* 2022; 127: #e2022JE007436. https://doi.org/10.1029/2022JE007436. DOI: https://doi.org/10.1029/2022JE007436.

323. Waddington CH. Towards a Theoretical Biology 4. Essays. Edinburgh, UK: Edinburgh University Press, 1972.

324. Wikipedia. Stuart Kaufman, https://en.wikipedia.org/wiki/Stuart_Kauffman (2024).

325. Waddington CH. *The Strategy of the Genes: A Discussion of Some Aspects of Theoretical Biology*. London, UK: George Allen & Unwin, 1957.

326. Allen M. Compelled by the diagram: Thinking through C. H. Waddington's epigenetic landscape. *Contemporaneity* 2015; 4: DOI 10.5195/contemp.2015.5143

327. Wikipedia. C. H. Waddington, https://en.wikipedia.org/wiki/C._H._Waddington (2024).

328. Waddington CH. *Principles of Embryology*. Macmillan, 1956.

329. Waddington CH. *Behind Appearance: A Study of the Relations Between Painting and the Natural Sciences in this Century*. MIT Press, 1970.

330. Flickinger RA. *Developmental Biology*. W.C. Brown Company, 1969.

331. Martin CC and Gordon R. Differentiation trees, a junk DNA molecular clock, and the evolution of neoteny in salamanders. *J Evol Biol* 1995; 8: 339–354.

332. Alicea B and Gordon R. Quantifying mosaic development: Towards an evo-devo postmodern synthesis of the evolution of development via differentiation trees of embryos [invited]. *Biology (Basel)* 2016; 5: #33.

333. Gordon R and Björklund NK. How to observe surface contraction waves on axolotl embryos. *Int J Dev Biol* 1996; 40: 913–914.

334. Gordon R and Gordon NK. The differentiation code. *BioSystems* 2019; 184. DOI: 10.1016/j.biosystems.2019.104013.

335. Gordon R and Stone R. A short tutorial on the Janus-faced logic of differentiation waves and differentiation trees and their evolution. *BioSystems* 2021; 205: #104414. DOI: 10.1016/j.biosystems.2021.104414.

336. Gordon R. Are we on the cusp of a new paradigm for biology? The illogic of molecular developmental biology versus Janus-faced control of embryogenesis via differentiation waves. *BioSystems* 2021. #104367.

337. Ahmad B and Gordon R. Is it a Janus-faced world after all? Physics is not reductionist [Chapter 3]. In: Sharov A and Mikhailovsky G (eds) *POEM, Pathways to the Origin and Evolution of Meanings in the Universe [Volume in the series: Astrobiology Perspectives on Life of the Universe book series (Series editors: Richard Gordon & Joseph Seckbach, Wiley-Scrivener)]*. Beverly, Massachusetts, USA: Wiley-Scrivener, 2023, pp. 55–70.

338. Igamberdiev AU and Gordon R. Macroevolution, differentiation trees, and the growth of coding systems. *BioSystems* 2023; 234: #105044. DOI: https://doi.org/10.1016/j.biosystems.2023.105044.

339. Björklund NK and Gordon R. Nuclear state splitting: A working model for the mechanochemical coupling of differentiation waves to master genes. *Russian Journal of Developmental Biology* 1993; 24: 79–95.

340. Santella L, Gordon R, Chen Z and Tuszynski J. Editorial: Waves in fertilization, cell division and embryogenesis. *Biosystems* 2021; 210: 2. Editorial Material. DOI: 10.1016/j.biosystems.2021.104560.

341. Chan Z, Santella L, Tuszynski J and Gordon R. Editorial: Waves in fertilization, cell division and embryogenesis. *BioSystems* 2020; 210: #104560.

342. Gordon R and Björklund NK. A comparison of cytoskeletal based differentiation waves in plants and animals. In: Baskin T and Rogers J (eds) *Sixteenth Annual Symposium, 1997, Current Topics in Plant Biochemistry and Molecular Biology, "Signs and Roadways: Protein Traffic and the Cytoskeleton", ER, Golgi, Actin, Vacuoles, Microtubules, Vesicular Traffic, Organelle Movements, Program/Abstracts*. Columbia, Missouri, USA: Interdisciplinary Plant Group, University of Missouri, 1997, pp. 84.

343. Library of Conciousness. Conrad Hal Waddington, https://www.organism.earth/library/author/conrad-waddington (2024).

344. Gordon R. Artifacts in reconstructions made from a few projections. In: Fu KS (ed) *Proceedings of the First International Joint Conference*

on Pattern Recognition, Oct 30 to Nov 1, 1973, Washington, D C. Northridge, California: IEEE Computer Society, 1973, pp. 275–285.

345. Greffier J, Si-Mohamed S, Frandon J, Loisy M, de Oliveira F, Beregi JP and Dabli D. Impact of an artificial intelligence deep-learning reconstruction algorithm for CT on image quality and potential dose reduction: A phantom study. *Med Phys* 2022; 49: 5052–5063. 2022/06/14. DOI: 10.1002/mp.15807.

346. Singh R, Wu WW, Wang G and Kalra MK. Artificial intelligence in image reconstruction: The change is here. *Physica Medica-European Journal of Medical Physics* 2020; 79: 113–125. Review. DOI: 10.1016/j. ejmp.2020.11.012.

347. Willemink MJ and Noël PB. The evolution of image reconstruction for CT from filtered back projection to artificial intelligence. *Eur Radiol* 2019; 29: 2185–2195. DOI: 10.1007/s00330-018-5810-7.

348. Gordon R and Poulin BJ. There is but one journal: The scientific literature. *PLoS Med* 2008; 5: #e201.

349. Wang C and Gordon R. How wrong could we be? A new way to solve underdetermined linear equations, illustrated via computed tomography *Journal of Applied and Numerical Optimization* 2024; 6: 351–370.

350. Martin D, Thulasiraman P and Gordon R. Local independence in computed tomography as a basis for parallel computing. *Technology in Cancer Research and Treatment* 2005; 4: 187–188.

351. Gordon R and Drum RW. An alternative view of morphogenesis. *Neurosciences Research Program Bulletin* 1974; 12: 44–45.

352. Babenko I, Kröger N and Friedrich BM. Mechanism of branching morphogenesis inspired by diatom silica formation. *Proceedings of the National Academy of Sciences* 2024; 121: #e2309518121. DOI: 10.1073/ pnas.2309518121.

353. Turing AM. The chemical basis of morphogenesis. *Philosophical Transactions of the Royal Society of London Series B, Biological Sciences* 1952; B237: 37–72.

354. Gordon R. Part Three: The reverse engineering road to computing life. Chapter 10: Walking the tightrope: The dilemmas of hierarchical instabilities in Turing's morphogenesis [Chapter 10, invited]. In: Cooper SB

and Hodges A (eds) *The Once and Future Turing: Computing the World*. Cambridge: Cambridge University Press, 2016, pp. 144–159.

355. Gordon NK, Chen Z, Gordon R and Zou YT. French flag gradients and Turing reaction-diffusion versus differentiation waves as models of morphogenesis. *BioSystems* 2020; 196. DOI: 10.1016/j.biosystems.2020.104169.

356. Paulos JA. *Innumeracy: Mathematical Illiteracy and Its Consequences*. Farrar, Straus and Giroux, 2011.

357. Drum RW, Pankratz HS and Stoermer EF. Diatomeenschalen im Elektronenmikroskopischen Bild. Teil VI. Electron Microscopy of Diatom Cells. In: Helmcke JG and Krieger W (eds). Lehre: J. Cramer, 1966.

358. Drum RW and Gordon R. Star Trek replicators and diatom nanotechnology. *Trends Biotechnol* 2003; 21: 325–328.

359. Mountain Rose Herbs. The Free Herbalism Project featuring Ryan Drum!, https://blog.mountainroseherbs.com/free-herbalism-project-featuring-ryan-drum (2015).

360. Gordon R. The chemical basis for diatom morphogenesis: instabilities in diffusion-limited amorphous precipitation generate space filling branch patterns [abstract]. *Fed Proc* 1981; 40: 827.

361. Gordon R. Instabilities in Turing's Morphogenesis. *The Once and Future Turing* 2016: 144.

362. Wikipedia. Alan Turing, https://en.wikipedia.org/wiki/Alan_Turing (2024).

363. Rushforth SJ, Edlund MB, Spaulding SA and Stoermer EF. The Reimer Diatom Herbarium: An important resource for teaching and research. *Proc Acad Nat Sci Phila* 2010; 160: 13–20. DOI: 10.1635/053.160.0103.

364. Gopal D, Chakrabarti S, Venkata D, Keshav S, Gordon R and Mazumder N. Recent insights into the ultrastructure of diatoms using scanning and transmission electron-microscopy [Chapter 3]. In: Mazumder N and Gordon R (eds) *Diatom Microscopy [DIMI, Volume in the series: Diatoms: Biology & Applications, series editors: Richard Gordon & Joseph Seckbach]*. Beverly, MA, USA: Wiley-Scrivener, 2022, pp. 57–80.

365. Pickett-Heaps JD and Pickett-Heaps J. Life in Glass Houses, http://www.limnolfwbiol.com/index.php/LFWB/article/view/801 (2023).

366. Aram L, Haan Dd, Varsano N, Gilchrist JB, Heintze C, Rotkopf R, Rechav K, Elad N, Kröger N and Gal A. Intracellular morphogenesis of diatom silica is guided by local variations in membrane curvature. *Nat Commun* 2024; 15: #7888. DOI: 10.1038/s41467-024-52211-x.

367. Taylor DL. Robert Day Allen (1927–1986). *Nature* 1986; 321: 647–647. DOI: 10.1038/321647a0.

368. Collings DA, Carter CN, Rink JC, Scott AC, Wyatt SE and Allen NS. Plant nuclei can contain extensive grooves and invaginations. *Plant Cell* 2000; 12: 2425–2440.

369. Kamiya N. PED-22125312 Y2024/M10/D17 Optical approaches to the dynamics of cellular motility: Marine biological laboratory centenary symposium, a symposium in honor of Robert D. Allen. *Cell Motility and the Cytoskeleton* 1988; 10: 11–12. Editorial Material. DOI: 10.1002/cm.970100104.

370. Wolpert L. Positional information and the spatial pattern of cellular differentiation. *J Theor Biol* 1969; 25: 1–47.

371. Gordon R and Björklund NK. Contraction and expansion waves cause cell differentiation in embryos: Mechanics of the cell state splitter [PowerPoint, CD]. *International Symposium on Muscular Contraction and Cell Movements, Colima, Mexico, January 20–25, 2005*. Colima, Mexico: University of Colima, 2005.

372. Gordon NK and Gordon R. Differentiation waves as the source of the 4 rules. Comment on: Four simple rules that are sufficient to generate the mammalian blastocyst. *PLoS Biology* 2017; 15: #e2000737.

373. Driesch HAE. *The Science and Philosophy of the Organism*. London: A. & C. Black, 1908.

374. Summerbell D and Wolpert L. Precision of development in chick limb morphogenesis. *Nature* 1973; 244: 228–230. DOI: 10.1038/244228a0.

375. Beloussov LV and Gordon R. Two ways for interpreting Driesch's law: "Positional information" and morphogenetic fields. *BioSystems* 2018; 173: 7–9. Editorial Material. DOI: 10.1016/j.biosystems.2018.09.005.

376. Gordon R and Beloussov LV. Special Session: Morphomechanics of the Embryo and Genome + Artificial Life -> Embryonics [English].

In: Mange D, Teuscher C, Holcombe M, *et al.* (eds) *IPCAT2003 5th International Workshop on Information Processing in Cells and Tissues, Swiss Federal Institute of Technology, Lausanne, Switzerland, School of Computer and Communication Sciences, Logic Systems Laboratory, September 8–11, 2003 [flyer]*. Lausanne, Switzerland: Swiss Federal Institute of Technology, 2003.

377. Gordon R. Making waves: The paradigms of developmental biology and their impact on artificial life and embryonics. *Cybernetics & Systems* 2001; 32: 443–458.

378. Gordon R. Mechanics in embryogenesis and embryonics: Prime mover or epiphenomenon? *Int J Dev Biol* 2003; 50: 245–253.

379. Gordon R. Make robots that build themselves like embryos. In: Bowermaster P and Gordon S (eds) *Visions for a World Transformed: 99 Ideas for Making the World a Better Place — Starting Right Now.* 2017, pp. 84–85.

380. Beloussov LV and Lipchinsky A. *Morphomechanics of Development.* Cham: Springer, 2015.

381. Gordon R. The publications of embryologist Lev V. Beloussov. *BioSystems* 2018; 173: 4–6. DOI: 10.1016/j.biosystems.2018.10.013.

382. Cohn SA, Manoylov KM and Gordon R. Dedication to Jeremy D. Pickett-Heaps, in memoriam, 1940–2021. In: Cohn SA, Manoylov KM and Gordon R (eds) *Diatom Gliding Motility [DIGM, Volume in the series: Diatoms: Biology & Applications, series editors: Richard Gordon & Joseph Seckbach].* Beverly, MA, USA: Wiley-Scrivener, 2021, pp. v-xviii.

383. Hildebrand M and Gordon R. The effect of the silica cell wall on diatom transport and metabolism/Publications by and about Mark Hildebrand [Chapter 10]. In: Annenkov V, Seckbach J and Gordon R (eds) *Diatom Morphogenesis [DIMO, Volume in the series: Diatoms: Biology & Applications, series editors: Richard Gordon & Joseph Seckbach].* Beverly, MA, USA: Wiley-Scrivener, 2021, pp. 253–260.

384. Gordon R. A tutorial on ART (Algebraic Reconstruction Techniques) [Erratum in Eq. 18: max, not min]. *IEEE Trans Nucl Sci* 1974; NS-21: 78–93, 95.

385. Gordon R and Lauterbur PC. Introduction to the session on experimental aspects of reconstruction from projections. In: Marr RB (ed)

Techniques of Three-Dimensional Reconstruction, Proceedings of an International Workshop. Upton, New York: Brookhaven National Laboratory, 1974, pp. 17–19.

386. National Inventors Hall of Fame. Paul Christian Lauterbur, Magnetic Resonance Imaging, https://www.invent.org/sites/default/files/styles/inductee_media/public/inductees/346-master_1.jpg?h=754009f3 (2024).

387. Nobel Prize Outreach. Paul C. Lauterbur Facts, https://www.nobelprize.org/prizes/medicine/2003/lauterbur/facts/ (2003).

388. Wikipedia. Paul Lauterbur, https://en.wikipedia.org/wiki/Paul_Lauterbur (2024).

389. Lauterbur PC. Image formation by induced interactions: Examples employing nuclear magnetic resonance. *Nature* 1973; 242: 191–192.

390. Dawson MJ. *Paul Lauterbur and the Invention of MRI*. Cambridge, Massachusetts, USA: MIT Press, 2013.

391. Lauterbur PC. Demystifying biology: Did life begin as a complex system? *Complexity* 2005; 11: 30–35. DOI: 10.1002/cplx.20097.

392. Kinsinger LA, Lauterbur PC and Baudry J. Modeling the interactions of a silica surface embedded with a dipeptide imprint. *Abstracts of Papers of the American Chemical Society* 2006; 231: 1084-CHED.

393. Lauterbur PC, Dawson MJ and Girolami G. The spontaneous development of biology from chemistry. *Astrobiology* 2008; 8: 3–8. DOI: 10.1089/ast.2008.1109.

394. Feng L, Pamidighantam B and Lauterbur PC. Microwave-assisted sol-gel synthesis for molecular imprinting. *Anal Bioanal Chem* 2010; 396: 1607–1612. DOI: 10.1007/s00216-009-3311-x.

395. Hazen RM, Liu X-M, Downs RT, Golden J, Pires AJ, Grew ES, Hystad G, Estrada C and Sverjensky DA. Mineral evolution: Episodic metallogenesis, the supercontinent cycle, and the coevolving geosphere and biosphere [Special Publication 18]. In: Kelley KD and Golden HC (eds) *Building Exploration Capability for the 21st Century*. Littleton, Colorado, USA: Society of Economic Geologists, 2014, pp. 1–15.

396. Herman GT, Liu WH, Rowland S and Walker A. Synchronization of growing cellular arrays. *Information and Control* 1974; 25: 103–122.

397. Descouens D. File:Cône textileI.jpg, https://commons.wikimedia.org/wiki/File:C%C3%B4ne_textileI.jpg (2009).

398. Waddington CH and Cowe RJ. Computer simulation of a mulluscan pigmentation pattern. *J Theor Biol* 1969; 25: 219–225. DOI: 10.1016/s0022-5193(69)80060-3.

399. Herman GT. SNARK93: A programming system for 2-D image reconstruction from projections for the UNIX/Sun environment, https://foss.commons.gc.cuny.edu/snark09/ (2001).

400. Lindenmayer A. Mathematical models for cellular interactions in development II. Simple and branching filaments with two-sided inputs. *J Theor Biol* 1968; 18: 300–315. DOI: https://doi.org/10.1016/0022-5193(68)90080-5.

401. Wikipedia. Aristid Lindenmayer, https://en.wikipedia.org/wiki/Aristid_Lindenmayer (2023).

402. Wolfram S. *A New Kind of Science*. Champaign, Illinois: Wolfram Media, 2001.

403. Gordon R and Herman GT. Three-dimensional reconstruction from projections: A review of algorithms. *International Review of Cytology* 1974; 38: 111–151.

404. Gordon R and Herman GT. Three-dimensional reconstruction from projections: a review of algorithms. *Proc SPIE* 1975; 47: 2–14.

405. Gordon R. Digest of Technical Papers, Topical Meeting on Image Processing for 2-D and 3-D Reconstruction from Projections: Theory and Practice in Medicine and the Physical Sciences. Washington, D.C.: Optical Society of America, 1975.

406. Gordon R. *Digest of Technical Papers, Topical Meeting on Image Processing for 2-D and 3-D Reconstruction from Projections: Theory and Practice in Medicine and the Physical Sciences, Post-Deadline Papers*. Washington, D.C.: Optical Society of America, 1975.

407. Gordon R, Herman GT and Johnson SA. Image reconstruction from projections. *SciAm* 1975; 233: cover, 12, 56–61, 64–68, 139.

408. Gordon R. One man's noise is another man's data: The ARTIST algorithm for positron tomography [postdeadline paper, 2pp.]. *Topical Meeting on Signal Recovery and Synthesis with Incomplete Information and Partial Constraints*. Washington, D.C.: Optical Society of America, 1983.

409. Gordon R. Maximal use of single photons and particles in reconstruction from projections by ARTIST, Algebraic Reconstruction Techniques Intended for Storage Tubes. In: Gordon R (ed) *Technical Digest, Topical Meeting on Image Processing for 2-D and 3-D Reconstruction from Projections: Theory and Practice in Medicine and the Physical Sciences.* Washington, D.C.: Optical Society of America, 1975, pp. #TuC4.

410. Wikipedia. Greedy algorithm, http://en.wikipedia.org/wiki/Greedy_algorithm (2023).

411. Gordon R. The ARTIST algorithm for high resolution, low dose positron tomography. In: Menon D and Filipow LJ (eds) *Positron Emission Tomography, MARIA Design Symposium.* Edmonton: Medical Accelerator Research Institute in Alberta, Department of Applied Sciences in Medicine, University of Alberta, 1982, pp. 182.

412. Pawlak B and Gordon R. Density estimation for positron emission tomography. *Technol Cancer Res Treat* 2005; 4: 131–142.

413. Gordon R, Yeoh TW, Keen KJ and Braun WJ. Squeezing pictures out of photons in positron tomography and x-ray computed tomography. In: El-Shaarawi AH (ed) *XXVIIth Annual Meeting of the Statistical Society of Canada, June 6–9, 1999, University of Regina, Regina, Saskatchewan, Program with Abstracts.* Regina: University of Regina, 1999, pp. 70.

414. Gordon R and Bender R. Three-dimensional algebraic reconstruction techniques: A preliminary course. *J Theor Biol* 1971; 32: 217.

415. Bracewell RN and Riddle AC. Inversion of fan-beam scans in radio astronomy. *Astrophysical Journal* 1967; 150: 427–434. DOI: 10.1086/149346.

416. Wikipedia. Ronald N. Bracewell, https://en.wikipedia.org/wiki/Ronald_N._Bracewell (2024).

417. Verly JG and Bracewell RN. Image reconstruction from strip integrals in computer aided x-ray tomography. *Phys Med Biol* 1980; 25: 772.

418. Verly JG and Bracewell RN. Blurring in tomograms made with x-ray-beams of finite width. *J Comput Assist Tomogr* 1979; 3: 662–678. DOI: 10.1097/00004728-197910000-00017.

419. Bracewell RN and Wernecke SJ. Image reconstruction over a finite field of view. *J Opt Soc Am* 1975; 65: 1342–1346. DOI: 10.1364/josa.65.001342.

420. Bremermann HJ. Complexity and transcomputability. *The Encyclopedia of Ignorance I Physical Sciences*. Pergaon Press, 1977, pp. 167–174.

421. Gordon R. Questions of Uniqueness and Resolution in Reconstruction from Projections (Book Review). *Phys Today* 1979; 32: 52–56.

422. Gordon R. A bibliography on reconstruction from projections. *Digest of Technical Papers, Topical Meeting on Image Processing for 2-D and 3-D Reconstruction from Projections: Theory and Practice in Medicine and the Physical Sciences, August 4–7, 1975, Stanford University.* Washington, D. C.: Optical Society of America, 1975.

423. Gordon R. Reconstruction from projections in medicine and astronomy. In: van Schoonveld C (ed) *Image Formation from Coherence Functions in Astronomy, Proceedings of the IAU Colloquium No 49 on the Formation of Images from Spatial Coherence Functions in Astronomy, Held at Groningen, The Netherlands, 10–12 August 1978*. Dordrecht: D. Reidel Publishing Company, 1978, pp. 317–325.

424. Gordon R. Delegate. *Conference on the Economic, Scientific and Technological Benefits of the Canadian Long Base Line Array to Canada.* Winnipeg, 1983.

425. Som S. Believe it or not, we can (kind of) measure the air pressure of early Earth! Blue-green algae's response to different nitrogen pressures may tell us what fossils to look for and what those fossils tell us about ancient air pressure, https://sciworthy.com/believe-it-or-not-we-can-kind-of-measure-the-air-temperature-of-early-earth/ (2019).

426. Silverman SN, Kopf S, Som SM, Bebout BM and Gordon R. Morphological and isotopic changes of heterocystous cyanobacteria in response to N_2 partial pressure. *Geobiology* 2017, 1, 60–75.

427. Silverman SN, Gordon R, Bebout BM and Som SM. Morphological and isotopic changes of heterocystous cyanobacteria in response to N_2 partial pressure. *Geobiology* 2018: https://doi.org/10.1111/gbi.12312.

428. Gould RG, Couch JL, Napel S, Peschmann KR, Rand RE and Boyd DP. Performance characteristics of an ultrafast scanning electron beam CT scanner. *Med Phys* 1983; 10: 526.

429. Lipton MJ, Higgins CB, Farmer D and Boyd DP. Cardiac imaging with a high-speed cine-CT scanner: Preliminary-results. *Radiology* 1984; 152: 579–582. DOI: 10.1148/radiology.152.3.6540463.

430. Svahn TM, Gordon R, Ast JC, Riffel J and Hartbauer M. Comparison of photon-counting and flat-panel digital mammography for the purpose of 3D imaging using a novel image processing method. *Radiat Prot Dosim* 2021: 1–8.

431. Boyd D. Douglas and Inyoung Boyd, https://www.dboyd.com/ (2024).

432. Wikipedia. Godfrey Hounsfield, https://en.wikipedia.org/wiki/Godfrey_Hounsfield (2024).

433. Gordon R. Dose reduction in computerized tomography [Guest Editorial]. *Invest Radiol* 1976; 111: 508–517.

434. Gordon R, Swindell W and Barrett HH. Higher-resolution tomography. *Phys Today* 1978; 31: 46.

435. Gordon R. Dose reduction in computed tomography. *J Comput Assist Tomogr* 1977; 1: 251.

436. Gordon R. Dose reduction in computed tomography. In: Di Chiro G and Brooks RA (eds) *International Symposium on Computer Assisted Tomography in Nontumoral Diseases of the Brain, Spinal Cord and Eye, Book of Abstracts*. Bethesda: National Institutes of Health, 1976, pp. 2.

437. Gordon R. Low dose computer imaging, computed tomography and teleradiology for dentistry. *J Dental Res* 1984; 63: 523.

438. Gordon R. Feedback control of exposure geometry in dental radiography workshop, University of Connecticut, 16 May 1978. *Appl Optics* 1979; 18: 1769, 1834.

439. Gordon R. Conceptual summary and conclusions. In: Webber RL, (ed.). *Feedback Control of Exposure Geometry in Dental Radiography [Publ No 80–1954]*. Bethesda: U.S. National Institutes of Health, 1979, p. 31–33.

440. The NIH Record. Webber, 25-Year Corps Vet, Retires. *The NIH Record* 1989; XLI: 9.

441. Kalos MH, Davis SA, Mittelman PS and Mastras P. *Conceptual Design of a Vapor Fraction Instrument [http://www.osti.gov/energycitations/product.biblio.jsp?query_id=0&page=0&osti_id=4837780]*. White Plains, NY: Nuclear Development Corporation of America, 1961, p. 31.

442. Cormack AM. Representation of a function by its line integrals, with some radiological applications. *J Appl Phys* 1963; 34: 2722–2727. DOI: 10.1063/1.1729798.

443. Cormack AM. Representation of a function by its line integrals, with some radiological applications. II. *J Appl Phys* 2004; 35: 2908–2913. DOI: 10.1063/1.1713127.

444. Cormack AM. Representation of a function by its line integrals, with some radiological applications. II. *J Appl Phys* 1964; 35: 2908–2913. DOI: 10.1063/1.1713127.

445. Cormack AM. Reconstruction of densities from their projections, with applications in radiological physics. *Phys Med Biol* 1973; 18: 195–207. DOI: 10.1088/0031-9155/18/2/003.

446. Nobel Prize in Physiology or Medicine. Allan M. Cormack Biographical, https://www.nobelprize.org/prizes/medicine/1979/cormack/biographical/ (1979).

447. Gordon R and Kane J. Three-dimensional reconstruction: The state of the "ART". *IV International Biophysics Congress, Moscow, Abstracts of Contributed Papers* 1972; 2: 37.

448. Simonov VI and Feigin LA. Boris Konstantinovich Vainshtein 1921–1996 — Obituary. *Acta Crystallogr Sect A* 1997; 53: 531–534. Item About an Individual. DOI: 10.1107/s010876739700593x.

449. Lemkin P, Shapiro B, Gordon R and Lipkin LE. *PROC10 — An Image Processing Program for the PDP-10: Operation and Description [NCI/IP Technical Report #8].* Bethesda, Maryland, USA: U.S. National Cancer Institute Image Processing Unit, NIH, 1976, p. 46.

450. Lemkin P. The National Cancer Institute Real Time Picture Processor, https://history.nih.gov/museum/practitioners/RTPP/index.html; http://lemkingroup.com/HistoryOfRTPPproject/index.html (2011).

451. Lemkin P. The Lemkin Group, http://lemkingroup.com/ (2024).

452. Anonymous. Dr. Coulombre Named NICHD Assoc. Director For Intramural Research *The NIH Record* 1967; XIX: 1,7.

453. Coulombre AJ. Cytology of the developing eye. *International Review of Cytology-A Survey of Cell Biology* 1961; 11: 161–194. DOI: 10.1016/s0074-7696(08)62715-1.

454. Wikipedia. Azriel Rosenfeld, https://en.wikipedia.org/wiki/Azriel_Rosenfeld (2024).

455. Anonymous. Dr. Prewitt Honored for Computer Contributions to Medicine. *NIH Record* 1981; 33: 9.

456. Gordon R. Computational embryology of the vertebrate nervous system. In: Geisow MJ and Barrett AN (eds) *Computing in Biological Science*. Amsterdam: Elsevier/North-Holland, 1983, pp. 23–70.

457. Shaw PJ and Rawlins DJ. The point spread function of a confocal microscope: Its measurement and use in deconvolution of 3-D data. *J Microsc* 1991; 163: 151–165. Article. DOI: 10.1111/j.1365-2818.1991. tb03168.x.

458. Gordon R, Silver L and Rigel DS. Halftone graphics on computer terminals with storage display tubes. *Proc Soc Information Display* 1976; 17: 78–84.

459. Wikipedia. Storage tube, https://en.wikipedia.org/wiki/Storage_tube (2023).

460. Bender R and Rowe Jr JE. Microdensitometers and data processing for electron microscopy. *Proceedings, Twenty-ninth Annual Meeting, Electron Microscopy Society of America, Boston, Massachusetts, August 9, 10, 11, 12 and 13, 1971*. Baton Rouge: Claitor's Publishing Division, 1971, pp. 98–99.

461. Feldmann R, Bing D, Potter M, Mainhart C, Furie B, Furie B and Caporale L. Part II. Computer-assisted macromolecular structure generation: Extension of existing information: On the construction of computer models of proteins by the extension of crystallographic structures. *Ann NY Acad Sci* 2006; 439: 12–43. DOI: 10.1111/j.1749-6632.1985. tb25787.x.

462. Jacobson AG and Gordon R. Changes in the shape of the developing vertebrate nervous system analyzed experimentally, mathematically and by computer simulation. *J Exp Zool* 1976; 197: 191–246.

463. Gordon R and Jacobson AG. The shaping of tissues in embryos. *SciAm* 1978; 238: 106–113, 160.

464. Wallingford JB and Sater AK. Commentary and tribute to Antone Jacobson: The pioneer of morphodynamics. *Dev Biol* 2019; 451: 97–133. 2019/06/24. DOI: 10.1016/j.ydbio.2019.04.019.

465. University of Texas at Austin. Antone Gardner Jacobson, https://w3. biosci.utexas.edu/experimentalembryology/ (2012).

466. De Luca T and Gordon R. A systems engineering approach to perception. *Toward a Science of Consciousness 2008, April 8–12, 2008, Tucson, Arizona*. Tucson: Center for Consciousness Studies, 2008.

467. Jacobson AG and Gordon R. Nature and origin of patterns of changes in cell shape in embryos. *Journal of Supramolecular Structure* 1976; 5: 371–380.

468. Smith SL, Whitmore EL, Jacobson AG and Gordon R. Three dimensional reconstruction of neurulation. *Proceedings of the Microscopical Society of Canada, Fourteenth Annual Meeting, 17–19 June 1987, University of Manitoba, Winnipeg, Manitoba.* Toronto: Microscopical Society of Canada, 1987, pp. 10–11.

469. Jacobson AG and Gordon R. Nature and origin of patterns of changes in cell shape in embryos. *Progress in Clinical Biological Research* 1977; 17: 323–332.

470. Jacobson AG. Some forces that shape the nervous system. *Zoon* 1978; 6: 13–21.

471. Markandeya V. Energy: Gasoline for Keeps? "Milking" algae could one day make petrol cars sustainable. *Scientific American India* 2009; 4: 19.

472. Scientific American Editors. Dr. No Money: The broken science funding system. Scientists spend too much time raising cash instead of doing experiments. *Scientific American Magazine* 2011; 304: http://www.scientificamerican.com/article.cfm?id=dr-no-money.

473. Gordon R and Hirsch HVB. Vision begins with direct reconstruction of the retinal image, how the brain sees and sfores pictures. In: Schallenberger H and Schrey H (eds) *Gegenstrom, für Helmut Hirsch zum Siebzigsten/Against the Stream, for Helmut Hirsch on His 70th Birthday.* Wuppertal: Peter Hammer Verlag GmbH, 1977, pp. 201–214.

474. Wikipedia. Karl H. Pribram, https://en.wikipedia.org/wiki/Karl_H._Pribram (2024).

475. Goleman D and Pribram K. Holographic memory. *Psychology Today* 1979; 12: 71–84.

476. Gordon R and Tweed DB. Quantitative reconstruction of visual cortex receptive fields. *University of Manitoba Medical Journal* 1983; 53: 75.

477. Gordon R. High-speed reconstruction of the finest details available in x-ray projections. In: Ter-Pogossian MM, Phelps ME, Brownell GL, *et al.* (eds) *Reconstruction Tomography in Diagnostic Radiology and Nuclear Medicine [Proceedings of a Workshop on Reconstruction Tomography*

in Diagnostic Radiology and Nuclear Medicine, San Juan, Puerto Rico April 17–19, 1975]. Baltimore: University Park Press, 1977, pp. 77–83.

478. Middleton NT and Harman MT. A preprocessor for geotomographic imaging of irregular geometric scans. *IEEE Trans Ind Appl* 1992; 28: 1148–1153. DOI: 10.1109/28.158841.

479. Gordon R. Computed tomography for rural Manitoba via teleradiology. In: MacEwan D and Wardrop D (eds) *Proceedings, Telehealth '79, October 23–24, 1979, Winnipeg.* Winnipeg: Manitoba Telephone System, 1979, pp. 69–75.

480. Rangaraj MR and Gordon R. Computed tomography for remote areas via teleradiology. *Proc SPIE* 1982; 318: 182–185.

481. Gordon R, Rangayyan RM, Wardrop DH and Beeman TM. Improving image quality in teleradiology and tele-computed tomography. *Proc IEEE Conference, Systems, Man and Cybernetics Society, Bombay, New Delhi, December 30 – January 7.* 1983, pp. 908–913.

482. Winnipeg Free Press. Dennis Hugh Wardrop, https://passages.winnipegfreepress.com/passage-details/id-207641/Dennis_Wardrop (2013).

483. Rangayyan RM and Gordon R. Proceedings IEEE WESCANEX '90, IEEE Western Canada Conference and Exhibition on Telecommunication for Health Care: Telemetry, Teleradiology, and Telemedicine [Proc. SPIE Volume 1355]. Bellingham, Washington: International Society for Optical Engineering, 1990.

484. Gordon R. The opportunity for teleradiology in Manitoba. *Computer Post* 1991; 1: 17, 20.

485. Gordon R and Rangayyan RM. Computed tomography for remote areas via digital teleradiology. *Proc 45th Annual Meeting of Canadian Assoc of Radiologists.* 1982, pp. 179.

486. Rangayyan RM and Gordon R. Expanding the dynamic range of x-ray videodensitometry for digital mammography and teleradiology. *Proc 45th Annual Meeting of Canadian Assoc of Radiologists.* 1982, pp. 189.

487. Rangayyan RM and Gordon R. Computed tomography from ordinary radiographs for teleradiology. *Med Phys* 1983; 10: 687–690. 1983/09/01.

488. Gordon R. The need for cross-fertilization between the fields of profile inversion and computed tomography. In: Best WG and Weselake SA (eds) *Remote Sensing Information, Seventh Canadian Symposium on*

Remote Sensing, Winnipeg, September 8 to 11, 1981. Ottawa: Canadian Remote Sensing Society of the Canadian Aeronautics and Space Institute, 1981, p. 107.

489. McDade IC and Llewellyn EJ. Satellite airglow limb tomography: Methods for recovering structured emission rates in the mesospheric airglow layer. *Can J Phys* 1993; 71: 552–563. DOI: 10.1139/p93-084.

490. Solomon SC, Hays PB and Abreu VJ. Tomographic inversion of satellite photometry. *Appl Optics* 1984; 23: 3409–3414. DOI: 10.1364/ao.23.003409.

491. Gordon R. PED-22121713 Rotating microscope for "LANDSAT" photography of vertebrate embryos. *Proc SPIE* 1982; 361: 48–52.

492. Gordon R and Rangaraj MR. The need for cross-fertilization between the fields of profile inversion and computed tomography. In: Best WG and Weselake SA (eds) *Proceedings of the Seventh Canadian Symposium on Remote Sensing, Winnipeg, September 8 to 11, 1981*. Ottawa: Canadian Remote Sensing Society of the Canadian Aeronautics and Space Institute, 1981, pp. 538–540.

493. Gordon R. A rotating microscope for observing all cells during neurulation. *University of Manitoba Medical Journal* 1983; 53: 79.

494. Crawford-Young SJ, Dittapongpitch S, Gordon R and Harrington KIS. Acquisition and reconstruction of 4D surfaces of axolotl embryos with the flipping stage robotic microscope. *BioSystems* 2018; 173: 214–220. DOI: 10.1016/j.biosystems.2018.10.006.

495. Gruwel MLH, Lerner B, Crawford-Young S and Gordon R. Early development of axolotl (*Ambystoma mexicanum*) embryos using MRI micro-imaging. *Manitoba Institute of Child Health 3rd Annual Research Day, October 11, 2007*. Winnipeg: Manitoba Institute of Child Health, 2007, pp. 19.

496. Gordon R. The future of computers in medicine. *University Manitoba Med J* 1983; 53: 27–31.

497. Blyden ER and Gordon R. Genomics, pharmacology and 3D imaging: Self-knowledge in the post-genomic era. *Biotechnology Focus* 2000; 3: 14, 16.

498. Blyden ER. Eggs could save the world from COVID-19, https:// medium.com/@eluem.blyden/eggs-could-save-the-world-from-covid-19-9b81c644d21f (2020).

499. Rosenbaum W, Syrotuik J and Gordon R. Random number generators for microcomputers. *Computer Programs Biomed* 1983; 16: 235–239.

500. Zhou X and Gordon R. Generation of noise in binary images. *Computer Vision, Graphics and Image Processing: Graphical Models and Image Processing* 1991; 53: 476–478.

501. Levina S and Gordon R. Methionine enkephalin-induced changes in pigmentation of zebrafish (Cyprinidae, *Brachydanio rerio*) and related species and varieties, measured videodensitometrically. I. Zebrafish. *Gen Comp Endocrinol* 1983; 51: 370–377.

502. King GM, Gordon R, Karmali K and Biberman LJ. A new method for the immobilisation of teleost embryos for time-lapse studies of development. *J Exp Zool* 1982; 220: 147–151.

503. Mummery MR. Does contact inhibition have a role in pigment cell interactions? [B.Sc.(Med.) Thesis, Advisor: Richard Gordon]. *Program and Proceedings, BSc (Med) Presentations, 1982*. Winnipeg: Faculty of Medicine, University of Manitoba, 1982, pp. 121–128.

504. King GM and Gordon R. In vivo migration of melanocytes in the zebrafish embryo (*Brachydanio rerio*). *Fed Proc* 1981; 40: 556.

505. Gordon R and King GM. Pattern formation by migrating melanocytes in teleost embryos. *Abstracts, The 40th Annual Symposium of The Society for Developmental Biology, Developmental Order: Its Origin and Regulation*. Boulder: University of Colorado, 1981, pp. 2.

506. Choudhri S. Electron microscopy of cell-cell contacts between migrating yolk sac melanocytes of the zebrafish, *Brachydanio rerio* [Supervisor: Richard Gordon]. *UM Program and Proceedings, BSc (Med) Presentations*. Winnipeg: Faculty of Medicine, University of Manitoba, 1984, pp. 71–78.

507. Gordon R and Levina SE. Methionine enkephalin-induced changes in pigmentation of zebrafish (Cyprinidae, *Brachydanio rerio*) and related species and varieties, measured videodensitometrically. II. Pearl and gold danios. *Gen Comp Endocrinol* 1983; 51: 378–383.

508. Rangayyan RM and Gordon R. Expanding the dynamic range of x-ray videodensitometry using ordinary image digitizing devices. *Appl Optics* 1984; 23: 3117–3120.

509. Dhawan AP, Gordon R and Rangayyan RM. Nevoscopy: Three-dimensional computed tomography for nevi and melanomas in situ by transillumination. *IEEE Trans Med Imaging* 1984; MI-3: 54–61.

510. Gordon R. Three dimensional computed tomography of nevi in situ by transillumination. *Anal Quant Cytol* 1983; 5: 208.

511. Rangayyan RM, Dhawan AP and Gordon R. Algorithms for limited-view computed tomography: a survey. *IEEE Int Conf on Computers, Systems and Signal Processing*. Bangalore, India: IEEE Press, 1984, pp. 1540–1545.

512. Dhawan AP, Gordon R and Rangayyan RM. Computed tomography by transillumination to detect early melanoma. *IEEE Trans Biomed Eng* 1984; BME-31: 574.

513. Dhawan AP, Rangayyan RM and Gordon R. Image restoration by two-dimensional deconvolution in limited-view reconstruction. In: Gordon R (ed) *Topical Meeting on Industrial Applications of Computed Tomography and NMR Imaging, Technical Digest, August 13–14, 1984, Hecla Island, Manitoba, Canada*. Washington, DC, USA: Optical Society of America, 1984, pp. TuA5-1 to TuA5-3.

514. Gordon R, Dhawan AP and Rangayyan RM. Image restoration in limited-view CT. *Radiology* 1984; 153: 369.

515. Dhawan AP, Gordon R, Lavelle CLB, Beeman T, Pond A and Carvalho V. Three-dimensional analysis at the microscopic level. In: Bance GN (ed) *Proceedings of the Microscopical Society of Canada, Twelfth Annual Meeting, 18–20 May 1985, University of New Brunswick, Fredericton, NB*. Toronto: Microscopical Society of Canada, 1985, pp. 34–35.

516. Dhawan AP, Rangayyan RM and Gordon R. Image restoration by Wiener deconvolution in limited-view computed tomography. *Appl Optics* 1985; 24: 4013-4020.

517. Gordon R, Dhawan AP and Rangayyan RM. Reply to "Comments on geometric deconvolution: a meta-algorithm for limited view computed tomography". *IEEE Trans Biomed Eng* 1985; 32: 242–244. Note.

518. Rangayyan RM, Dhawan AP and Gordon R. Algorithms for limited-view computed tomography: An annotated bibliography and a challenge. *Appl Optics* 1985; 24: 4000–4012.

519. Dhawan AP, Buelloni G and Gordon R. Correction to "Enhancement of mammographic features by optimal adaptive neighborhood image processing". *IEEE Trans Medical Imaging* 1986; MI-5: 120.

520. Dhawan AP, Buelloni G and Gordon R. Enhancement of mammographic features by optimal adaptive neighborhood image processing [+Correction: **MI-5**(2), 120]. *IEEE Trans Medical Imaging* 1986; MI-5: 8–15, 120.

521. Dhawan AP, Le Royer E and Gordon R. Adaptive neighborhood image processing for feature enhancement of film-mammograms. *IEEE Eighth Annual Conference of the Engineering in Medicine and Biology Society, Nov 7–10, Dallas, TX.* 1986, pp. 1096–1100.

522. Dhawan AP and Gordon R. Reply to comments on "Enhancement of mammographic features by optimal adaptive neighborhood image processing". *IEEE Trans Med Imaging* 1987; MI-6: 82–83.

523. Steenstrup S, Dhawan AP and Gordon R. Comments on "Enhancement of mammographic features by optimal adaptive neighborhood image-processing"/Authors' reply. *IEEE Trans Med Imaging* 1987; 6: 82–83.

524. IEEE. Atam Dhawan, PhD, NJIT, https://www.embs.org/sc/atam-dhawan-phd-njit/ (2024).

525. Kamm J and Nagy JG. Optimal Kronecker product approximation of block Toeplitz matrices. *SIAM J Matrix Anal Appl* 2000; 22: 155–172. DOI: 10.1137/s0895479898345540.

526. Dhawan AP, Rangayyan RM and Gordon R. Wiener filtering for deconvolution of geometric artifacts in limited-view image reconstruction. *Proc SPIE* 1984; 515: 168–172.

527. Soble P, Rangayyan RM and Gordon R. Quantitative and qualitative evaluation of geometric deconvolution of distortion in limited-view computed tomography. *IEEE Trans Biomed Eng* 1985; BME-32: 330–335.

528. Gordon R and Rangayyan RM. Geometric deconvolution: a meta-algorithm for limited view computed tomography. *IEEE Trans Biomed Eng* 1983; 30: 806–810.

529. Rangayyan RM and Gordon R. Geometric deconvolution of artifacts in limited view computed tomography. *Digest of Papers, Topical Meeting on Signal Recovery and Synthesis with Incomplete Information and Partial Constraints.* Washington, D.C.: Optical Society of America, 1983, pp. FA2-1 to FA2-4.

530. Bamler R. Comments on "Geometric deconvolution: A meta-algorithm for limited view computed-tomography". *IEEE Trans Biomed Eng* 1985; 32: 241–242. DOI: 10.1109/tbme.1985.325535.

531. Wang ZL, Cai JH, Guo W, Donnelley M, Parsons D and Lee I. Backprojection Wiener deconvolution for computed tomographic reconstruction. *PLoS One* 2018; 13. DOI: 10.1371/journal.pone.0207907.

532. Walsh TE and Kihm KD. Tomographic deconvolution of laser speckle photography applied for flame temperature measurement. In: Crowder JP (ed) *Flow Visualization VII, 7th International Symposium on Flow Visualization, SEATTLE, Washington, USA, September 11–14, 1995.* New York, NY, USA: Begell House, 1995, pp. 898–903.

533. Skoglund U, Öfverstedt LG, Burnett RM and Bricogne G. Maximum-entropy three-dimensional reconstruction with deconvolution of the contrast transfer function: A test application with adenovirus. *J Struct Biol* 1996; 117: 173–188. DOI: 10.1006/jsbi.1996.0081.

534. Mory C, Auvray V, Zhang B, Grass M, Schäfer D, Peyrin F, Rit S, Douek P and Boussel L. Deconvolution for limited-view streak artifacts removal: Improvements upon an existing approach. In: Yu B (ed) *2012 IEEE Nuclear Science Symposium and Medical Imaging Conference Record.* 2012, pp. 2370–2373.

535. Hojjatoleslami SA, Avanaki MRN and Podoleanu AG. Image quality improvement in optical coherence tomography using Lucy-Richardson deconvolution algorithm. *Appl Optics* 2013; 52: 5663–5670. DOI: 10.1364/ao.52.005663.

536. Zhang ZC, Liang XK, Dong X, Xie YQ and Cao GH. A sparse-view CT reconstruction method based on combination of DenseNet and deconvolution. *IEEE Trans Med Imaging* 2018; 37: 1407–1417. DOI: 10.1109/tmi.2018.2823338.

537. Sudhakar P, Langoju R, Agrawal U, Patil BD, Narayanan A, Chaugule V, Amilneni V, Cheerankal P and Das B. Self-supervised learning for CT deconvolution. *Proc SPIE* 2021; 11595. DOI: 10.1117/12.2581269.

538. Vlasov V and Konovalov A. Minimizing the number of views in few-view computed tomography: A deep learning approach. 2022: preprint. DOI: 10.1109/ICIEAM54945.2022.9787247.

539. Portegys T, Vlasov V, Gordon R, Crawford-Young SJ, Konovalov AB and McGrew S. Can skin cancer be detected by a handheld scanner?, https://experiment.com/projects/can-skin-cancer-be-detected-by-a-handheld-scanner (2020).

540. Gordon R and Coumans J. Combining multiple imaging techniques for in vivo pathology: A quantitative method for coupling new imaging modalities. *Med Phys* 1984; 11: 79–80.

541. Gordon R and Rangayyan RM. Feature enhancement of film mammograms using fixed and adaptive neighborhoods [+Correction: **23**(13), 2055]. *Appl Optics* 1984; 23: 560–564.

542. Gordon R and Rangayyan RM. Radiographic feature enhancement, information content, and dose reduction in mammography and cardiac angiography. *Proc 5th Annual Conf Frontiers of Engineering and Computing in Health Care, IEEE-EMBS*. 1983, pp. 161–165.

543. Rangayyan RM and Gordon R. Streak preventive image reconstruction with ART and adaptive filtering. *IEEE Trans Med Imaging* 1982; MI-1: 173–178. DOI: 10.1109/tmi.1982.4307569.

544. Gordon R and Rangaraj MR. Experiments on streak prevention in image reconstruction from a few views. *Proceedings of the 4th Biennial Conference of the Canadian Society for Computational Studies of Intelligence, University of Saskatchewan, Saskatoon, Saskatchewan, 17–19 May 1982*. Toronto: Canadian Society for Computational Studies of Intelligence, 1982, pp. 41–47.

545. Dhawan AP and Gordon R. Enhancement of mammographic features by optimal adaptive neighborhood image-processing — reply. *IEEE Trans Med Imaging* 1987; 6: 82–83. DOI: 10.1109/tmi.1987.4307802.

546. Gordon R and Rangaraj MR. Computed tomography from a few ordinary radiographs. *IEEE Trans Biomed Eng* 1982; BME-29: 626.

547. Gordon R. Topical Meeting on Industrial Applications of Computed Tomography and NMR Imaging. Washington, D.C.: Optical Society of America, 1984.

548. Gordon R. Industrial applications of computed tomography and NMR imaging: an OSA topical meeting. *Appl Optics* 1985; 24: 3948–3949.

549. Wikipedia. Allan MacLeod Cormack, https://en.wikipedia.org/wiki/Allan_MacLeod_Cormack (2024).

550. Gordon R. Tutorial on survey of non-medical applications of computed tomography. In: Gordon R (ed) *Topical Meeting on Industrial Applications of Computed Tomography and NMR Imaging.* Washington, D.C.: Optical Society of America, 1984, pp. MA1-1.

551. Pragher W. File:34. Tagung 1984 Mediziner; Verabschiedung H.-H. Kunckel, Allen Cormack — W134Nr.123006 — Willy Pragher (cropped).jpg, https://commons.wikimedia.org/wiki/File:34._Tagung_1984_Mediziner;_Verabschiedung_H.-H._Kunckel,_Allen_Cormack_-_W134Nr.123006_-_Willy_Pragher_(cropped).jpg (1984).

552. Gordon R. Toward robotic x-ray vision: New directions for computed tomography. *Appl Optics* 1985; 24: 4124–4133.

553. Smith KT, Solmon DC and Wagner SL. Practical and mathematical aspects of the problem of reconstructing objects from radiographs. *Bull Amer Math Soc* 1977; 83: 1227–1270.

554. Leahy JV, Smith KT and Solmon DC. *Uniqueness, Nonuniqueness and Inversion in the X-ray and Radon Problems.* CINESTAV-IPN, 1992.

555. Faridani A, Finch DV, Ritman EL and Smith KT. Local tomography .2. *SIAM J Appl Math* 1997; 57: 1095–1127. DOI: 10.1137/s0036139995286357.

556. Faridani A, Ritman EL and Smith KT. Local tomography. *SIAM J Appl Math* 1992; 52: 459–484.

557. Faridani A, Ritman EL and Smith KT. Examples of local tomography. *SIAM J Appl Math* 1992; 52: 1193–1198. DOI: 10.1137/0152070.

558. Jaman KA, Gordon R and Rangayyan RM. Display of 3D anisotropic images from limited-view computed tomograms. *Computer Vision, Graphics, Image Processing* 1985; 30: 345–361.

559. Gordon R and Chen Y. From statistical mechanics to molecular biology: A Festschrift for Terrell L. Hill. *Cell Biophys* 1988; 12.

560. Gordon R and Brodland GW. The cytoskeletal mechanics of brain morphogenesis. Cell state splitters cause primary neural induction. *Cell Biophys* 1987; 11: 177–238.

561. Brodland GW and Gordon R. Mechanical instability of the cytoskeleton as a basis for primary neural induction. In: Butler DL and Torzilli PA (eds) *1987 Biomechanics Symposium*. New York: American Society of Mechanical Engineers, 1987, pp. 295–298.

562. Brodland GW and Gordon R. Intermediate filaments may prevent buckling of compressively-loaded microtubules. *Journal of Biomechanical Engineering* 1990; 112: 319–321.

563. Brodland GW and Gordon R. Intermediate filaments may prevent buckling of compressively-loaded microtubules. In: Medley J (ed) *Proceedings of the 9th Annual Conference of the Canadian Biomaterials Society, July 21–22, 1988, Waterloo, Ontario, Canada*. Waterloo: University of Waterloo, 1988, pp. 43–44.

564. Brodland GW and Gordon R. Intermediate filaments may prevent buckling of compressively loaded microtubules. *J Biomech Eng-Trans ASME* 1990; 112: 319–321. Note. DOI: 10.1115/1.2891190.

565. Gordon R and Tiffany MA. Possible buckling phenomena in diatom morphogenesis. In: Seckbach J and Kociolek JP (eds) *Diatom World*. Dordrecht: Springer, 2011, pp. 245–271.

566. Pappas JL and Gordon R. Buckling: A geometric and biophysical multiscale feature of centric diatom valve morphogenesis [Chapter 8]. In: Annenkov V, Seckbach J and Gordon R (eds) *Diatom Morphogenesis [DIMO, Volume in the series: Diatoms: Biology & Applications, series editors: Richard Gordon & Joseph Seckbach]*. Beverly, MA, USA: Wiley-Scrivener, 2021, pp. 195–230.

567. Tiffany MA, Nagy SS and Gordon R. The buckling of diatom valves. In: Edlund MB and Spaulding SA (eds) *North American Diatom Symposium (NADS), September 23–27, 2009, Iowa Lakeside Laboratory, Milford, Iowa*. Milford, Iowa: Iowa Lakeside Laboratory, University of Iowa, 2009, pp. 37–38.

568. Flint RW, Gordon R, Martin CC and Brodland GW. Simulation of the inversion of amphibian eggs in a gravitational field using hollow glass spheres. In: Ubbels GA, Oser H and Guyenne TD (eds) *Microgravity as a Tool in Developmental Biology, Selected papers from a special ESA Symposium held during the 11th International Congress of the International Society of Developmental Biologists, August 20–25, 1989, Utrecht.* Paris Cedex: European Space Agency, 1989, pp. 81–83.

569. Brodland GW, Gordon R, Scott MJ, Björklund NK, Luchka KB, Martin CC, Matuga C, Globus M, Vethamany-Globus S and Shu D. Furrowing surface contraction wave coincident with primary neural induction in amphibian embryos. *J Morphol* 1994; 219: 131–142.

570. Brodland GW, Scott MJ, MacLean AF, Globus M, VethamanyGlobus S, Gordon R, Veldhuis JH and DelMaestro R. Morphogenetic movements during axolotl neural tube formation tracked by digital imaging. *Roux's Archives of Developmental Biology* 1996; 205: 311–318. DOI: 10.1007/bf00365809.

571. Stein MB and Gordon R. Epithelia as bubble rafts: A new method for measuring cell shape and intercellular adhesion in embryonic and other epithelia. *J Theor Biol* 1982; 97: 625–635.

572. Brodland GW, Veldhuis JH, Kim S, Perrone M, Mashburn D and Hutson MS. CellFIT: A cellular force-inference toolkit using curvilinear cell boundaries. *PLoS One* 2014; 9. DOI: 10.1371/journal.pone.0099116.

573. Stein MB. Social Phobia: Clinical and Research Perspectives, (1995).

574. Harauz G, Gordon R and van Heel M. Oblique sampling of projections for direct three dimensional reconstruction. *Computer Vision, Graphics, Image Processing* 1987; 38: 81–89.

575. Gordon R. A retaliatory role for algal projectiles, with implications for the mechanochemistry of diatom gliding motility. *J Theor Biol* 1987; 126: 419–436.

576. Gordon R. Explosive Biopropulsion of Microorganisms [slides, Queen's]. In: Rainbow M and Rival DE (eds) *Conference on Biopropulsion of Adaptive Systems, Queen's University Biological Station, Ontario, Canada, July 23–26, 2018.* Kingston, Ontario, Canada: Queen's University, 2018.

577. Chang R and Prakash M. Biophysical limits of ultrafast cellular motility. *bioRxiv* 2024: #2024.2008. 2022.609204.

578. Gordon R and Brodland GW. Neurulation. In: Armstrong JB and Malacinski GM (eds) *Developmental Biology of the Axolotl*. New York: Oxford University Press, 1989, pp. 62–71.

579. Gordon R. A review of the theories of vertebrate neurulation and their relationship to the mechanics of neural tube birth defects. *Journal of Embryology and Experimental Morphology* 1985; 89: 229–255.

580. Gordon R. There is no such thing as safe sex: Chance of AIDS infection increases with time. *Winnipeg Free Press* 1987: 7, March 7.

581. Conant M, Fischl M, Gordon R, Jaffe H, Johnson W, Kaplan M, Minuk G, Peterman T, Sprecher S and Stille W. This program may shock you; and save your life [Radio Program: Quirks & Quarks broadcast April 4, 1987, Host: Jay Ingram, Guest(s): Dr. Marcus Conant, Dr. Margaret Fischl, Dr. Richard Gordon, Dr. Harold Jaffe, Warren Johnson, Dr. Mark Kaplan, Dr. Gerald Minuk, Dr. Tom Peterman, Dr. Suzanne Sprecher, Dr. Wolfgang Stille], https://www.cbc.ca/player/play/audio/1.3334017 (1987).

582. Gordon R and Björklund NK. If semen were red: The spread of red dye from the tips of condoms during intercourse and its consequences for the AIDS epidemic. *1990 World Health Organization International Conference, Assessing AIDS Prevention, Montreux, Switzerland.* 1990.

583. Charumilind S, Jain SH and Rhatigan J. *HIV in Thailand: The 100% Condom Program.* Harvard Business Publishing, 2011.

584. Shilts R. *And the Band Played On: Politics, People, and the AIDS Epidemic.* St. Martin's Press, 1987.

585. Rappe AGS. A contraceptive [patent AU5943590A], https://patents. google.com/patent/AU5943590A/en?assignee=axel+rappe&scholar&o-q=axel+rappe (1989).

586. Moghadas SM, Gumel AB, McLeod RG and Gordon R. Could condoms stop the AIDS epidemic? *Journal of Theoretical Medicine (Computational and Mathematical Methods in Medicine)* 2003; 5: 171–181. DOI: https://doi.org/10.1080/10273660412331315147.

587. Hardy-Lee Funeral Home. John TYSON Obituary, https://www.legacy.com/ca/obituaries/therecord-waterloo/name/john-tyson-obituary?id=40014092 (2022).

588. Gordon R and Tyson JE. Fatal sexually transmitted diseases (FSTDs), such as AIDS, select for the evolution of monogamy and provide a model for background extinction. *J Biol Syst* 1993; 1: 1–26. DOI: DOI: 10.1142/S0218339093000252.

589. Gordon R. Letters: One reader's meat, another's poison. *Sci News* 1994; 145: 131.

590. Wikipedia. File:Gumel-May2022.jpg, https://commons.wikimedia.org/wiki/File:Gumel-May2022.jpg (2023).

591. Friesen MRP, Gordon R and McLeod RD. Exploring emergence within social systems with agent based models [Chapter 4, invited]. In: Adamatti DF, Dimuro GP and Coelho H (eds) *Interdisciplinary Applications of Agent-Based Social Simulation and Modeling*. Hershey, Pennsylvania, USA IGI Global, 2014, pp. 52–71.

592. McLeod RD, Laskowski M, Friesen MRP, Podaima B, Gordon R and Wu J. *Agent Based Models for Nosocomial Infections: Modeling of hospital-acquired infections within a hospital. Final Report*. Winnipeg: Internet Innovation Centre, Electrical & Computer Engineering, University of Manitoba, 2009.

593. IEEE Explore. Robert D. Mcleod, https://ieeexplore.ieee.org/author/37318569700 (2011).

594. Niazi W and Gordon R. Stop Afghan AIDS (SAA) [Poster THPE0254, Abstract A-011-0147-20149]. *AIDS 2006, XVI International AIDS Conference 13–18 August 2006, Toronto Canada*. 2006.

595. Smith? RJ and Gordon R. The OptAIDS project: towards global halting of HIV/AIDS [Preface]. *BMC Public Health* 2009; 9: S1 (5 pages). DOI: doi:10.1186/1471-2458-9-S1-S1.

596. Smith? RJ and Gordon R. The OptAIDS project: towards global halting of HIV/AIDS Introduction. *BMC Public Health* 2009; 9: doi:10.1186/1471-2458-1189-s1181-s1181. DOI: 10.1186/1471-2458-9-s1-s1.

597. Smith? RJ, Li J, Gordon R and Heffernan J. Can we spend our way out of the AIDS epidemic? A world halting AIDS model. *BMC*

Public Health 2009; 9: S15 (17 pages). DOI: doi:10.1186/1471-2458-9-S1-S15.

598. Gordon R, Björklund NK, Smith? RJ and Blyden ER. Halting HIV/AIDS with avatars and havatars: A virtual world approach to modelling epidemics. *BMC Public Health* 2009; 9: S13 (16 pages). DOI: doi:10.1186/1471-2458-9-S1-S13.

599. Gordon R. OPTAIDS: Can We Optimally Spend Our Way Out of the HIV/AIDS Epidemic? World AIDS as an Optimal Halting Problem. Welcome! OptAIDS Workshop (http://www.mitacs.ca/conferences/OptAIDS/) July 29, 2008, Toronto and Second Life® [PowerPoint presentation], http://slurl.com/secondlife/Owlet/238/73/99 (2008).

600. Gordon R. Inexpensive computed tomography for remote areas. *Proc SPIE* 1990; 1355: 184–188.

601. Björklund NK, Gordon R and Martin CC. Defects due to compression of amphibian embryos at primary neural induction. In: Gordon R (ed) *Symposium, Mechanics of the Cytoskeleton in Developmental Biology, Sponsored by the Canadian Society for Theoretical Biology, 34th Annual Meeting, Canadian Federation of Biological Societies, Program/Proceedings, Queen's University, Kingston, Ontario, June 9–11, 1991.* Ottawa: Canadian Federation of Biological Societies, 1991, pp. 118.

602. Beloussov LV and Ermakov AS. Artificially applied tensions normalize development of relaxed *Xenopus laevis* embryos [English]. *Russian Journal of Developmental Biology* 2001; 32: 236–241. DOI: 10.1023/a:1016719219011.

603. Jaffe LF and Etkin L. Developmental Physiology, Gordon Research Conference, https://www.grc.org/developmental-physiology-conference/1998/ (1998).

604. Gordon R. The wave that makes the brain. In: Jaffe LF and Etkin L (eds) *Gordon Research Conference on Developmental Physiology, August 2 – 7, 1998.* Plymouth State College: Gordon Research Conferences, 1998, pp. https://www.grc.org/developmental-physiology-conference/1998/.

605. Gordon R, Beloussov L and Meinhardt H. From observations to paradigms; the importance of theories and models. An interview with Hans Meinhardt. *Int J Dev Biol* 2006; 50: 103–111.

606. Igamberdiev AU, Beloussov LV and Gordon R. Editorial. Special Issue: Biological Morphogenesis: Theory and Computation. *BioSystems* 2012; 109: 241–242.

607. Beloussov LV and Gordon R. Developmental morphodynamics — bridging the gap between the genome and embryo physics — Preface. *Int J Dev Biol* 2006; 50. DOI: 10.1387/ijdb.052137lb.

608. Beloussov LV and Gordon R. Morphodynamics session. In: Podlubnaya Z (ed) *International Symposium "Biological Motility" dedicated to the memory of academician GM Frank (1904–1976), Pushchino, Moscow region, Russia, May 23 – June 1, 2004.* Moscow, Russia: Scientific Council on Problems of Biological Physics, Russian Academy of Sciences, 2004.

609. Beloussov LV and Gordon R. Preface. Developmental morphodynamics: bridging the gap between the genome and embryo physics [English]. *Int J Dev Biol* 2006; 50: 79–80. DOI: 10.1387/ijdb.052137lb.

610. Beloussov LV. Life of Alexander G. Gurwitsch and his relevant contribution to the theory of morphogenetic fields. *Int J Dev Biol* 1997; 41: 771–779.

611. Beloussov LV. Morphogenetic fields: Outlining the alternatives and enlarging the context [English]. *Rivista di Biologia — Biology Forum* 2001; 94: 219–235. 2001/11/13.

612. Björklund NK and Gordon R. Surface contraction and expansion waves correlated with differentiation in axolotl embryos. III. The shape of the fate map. In: Konopka AK and Salamon P (eds) *Extended Abstracts and Topics for Discussion, Open Problems of Computational Molecular Biology, Third International Workshop, Telluride Summer Research Center, Telluride, CO, July 11–25, 1993.* Telluride: Telluride Summer Research Center, 1993, pp. 3–8.

613. Wikipedia. Rupert Sheldrake, https://en.wikipedia.org/wiki/Rupert_Sheldrake (2024).

614. Beloussov L. Lev Beloussov, Lomonosov Moscow State University, https://www.researchgate.net/profile/Lev-Beloussov (2021).

615. Rickey DW, Gordon R and Huda W. On lifting the inherent limitations of positron emission tomography by using magnetic fields (MagPET). *Automedica* 1992; 14: 355–369.

616. University of Manitoba. Daniel Rickey, https://umanitoba.ca/science/directory/physics-and-astronomy/daniel-rickey (2024).

617. Eleftheriou A, Tsoumpas C, Bertolli O and Stiliaris E. Effect of the magnetic field on positron range using GATE for PET-MR [abstract]. *EJNMMI Phys* 2014; 1: A50. DOI: 10.1186/2197-7364-1-S1-A50.

618. Gordon R and Luchka KB. Soliton waves on the vertebrate embryo, an excitable medium. *Workshop on Models, Mathematics and Numerics of Excitable Media and Mini-Conference on Mathematical Biology, October 1–4, 1992.* Winnipeg: Institute of Industrial Mathematical Sciences, University of Manitoba, 1992, pp. 32–33.

619. Jiang X, Guan H and Gordon R. Contrast enhancement using "feature pixels" or "fixels" for pixel-independent image processing. *Radiology* 1992; 185: 391.

620. ResearchGate. Huaiqun Guan, https://www.researchgate.net/profile/Huaiqun-Guan (2013).

621. Guan H and Gordon R. A projection access order for speedy convergence of ART (Algebraic Reconstruction Technique): a multilevel scheme for computed tomography. *Phys Med Biol* 1994; 39: 2005–2022.

622. Guan H and Gordon R. Computed tomography using Algebraic Reconstruction Techniques (ARTs) with different projection access schemes: A comparison study under practical situations. *Phys Med Biol* 1996; 41: 1727–1743.

623. Guan H, Gordon R and Zhu Y. Combining various projection access schemes with the Algebraic Reconstruction Technique for low-contrast detection in computed tomography. *Phys Med Biol* 1998; 43: 2413–2421.

624. Gordon R and Björklund NK. Differentiation waves in *Drosophila* and the axolotl. In: Chernoff EAG, Duhon ST, Malacinski GM, *et al.* (eds) *International Workshop on the Molecular Biology of Axolotls and other Urodeles, Indianapolis, Indiana, USA, October 13–16, 1993.* Bloomington: Indiana University, 1993, pp. 15–16.

625. Alicea B, Portegys TE, Gordon D and Gordon R. Morphogenetic processes as data: Quantitative structure in the *Drosophila* eye imaginal disc. *BioSystems* 2018; 173: 256–265.

626. Wolff T. *Drosophila* third instar eye disc. In: Bate M and Martinez Arias A (eds) *The Development of Drosophila melanogaster*. Plainview, New York: Cold Spring Harbor Laboratory Press, 1993, poster in rear pocket.

627. Wikipedia. David Suzuki, https://en.wikipedia.org/wiki/David_Suzuki (2024).

628. Poodry CA, Hall L and Suzuki DT. Temperature-sensitive mutations in *Drosophila melanogaster*. Part XV. Developmental properties of *shibi-re^{ts1}:* A pleiotropic mutation affecting larval and adult locomotion and development. *Dev Biol* 1973; 32: 373–386.

629. Gordon R. Grant agencies versus the search for truth. *Accountability in Research: Policies and Quality Assurance* 1993; 2: 297–301.

630. Meikle M, Wilson E and Gordon R. *The Science Game: Interview of Richard Gordon (Quirks and Quarks) [Audio tape, February 15, 1992].* Toronto: Canadian Broadcasting Corp., 1992.

631. Blogvideopost and Gordon R. Age Distribution of NIH Principal Investigators and Medical School Faculty, http://www.youtube.com/watch?v=rL_J-Yl55K0 (2012).

632. Berezin AA and Gordon R. Smaller grants for more Canadians? *Nature* 1997; 386: 212.

633. Berezin AA, Gordon R and Hunter G. Anonymous peer review and the QWERTY effect. *Amer Physics Soc News* 1995; 4: 7.

634. Berezin AA and Gordon R. Research under-funding or mismanagement? *University Affairs* 1997; 38: 25.

635. Berezin AA, Gordon R, Laznicka P, Rangacharyulu C and Torres ML. Canadian research funding. *Canadian Association of University Teachers (CAUT) Bulletin* 1997: 8.

636. Forsdyke DR and Gordon R. CSTB Workshop: Alternatives to the Present Granting System. *34th Annual Meeting, Canadian Federation of Biological Societies, Program/Proceedings, Queen's University, Kingston, Ontario, June 9–11, 1991.* Ottawa: Canadian Federation of Biological Societies, 1991, pp. 12.

637. Gordon R and Poulin BJ. Cost of the NSERC science grant peer review system exceeds the cost of giving every qualified researcher a baseline grant. *Accountability in Research: Policies and Quality Assurance* 2009; 16: 1–28.

638. Gordon R and Poulin BJ. Indeed: Cost of the NSERC science grant peer review system exceeds the cost of giving every qualified researcher a baseline grant. *Accountability in Research: Policies and Quality Assurance* 2009; 16: 232–233.

639. Poulin BJ and Gordon R. How to organize science funding: the new Canadian Institutes for Health Research (CIHR), an opportunity to vastly increase innovation. *Canadian Public Policy* 2001; 27: 95–112.

640. Gordon R. Testimony on breast cancer and mammography. *Canadian House of Commons Proceedings, Sub-Committee on the Status of Women* 1992: 1–23.

641. Gordon R and Poulin BJ. Testimony at Standing Committee on Health, Canadian Parliament, December 1, 1999, on Bill C-13, An Act to establish the Canadian Institutes of Health Research, to repeal the Medical Research Council Act and to make consequential amendments to other Acts, The Minister of Health, http://www.parl.gc.ca/HousePublications/Publication.aspx?DocId=1039912&Language=E&Mode=1 (1999).

642. Poulin BJ and Gordon R. *Statement on Peer Review to Standing Committee on Industry, Science and Technology, Parliament of Canada, November 29, 2001 [http://www.parl.gc.ca/InfoComDoc/37/1/INST/Meetings/Evidence/instev58-e.htm]*. Ottawa: Parliament of Canada, 2001.

643. Gordon R. Alternative reviews. *University Affairs (Association of Universities and Colleges of Canada)* 1993; 34: 26.

644. Berezin AA, Gordon R and Hunter G. Forum: Lifting the pernicious veil of secrecy; peer reviewers should shed the mask of anonymity argue three Canadian academics. *New Scientist* 1995; 145: 46–47.

645. Ferguson C, Marcus A, Oransky I and Gordon R. The peer-review scam. *Nature* 2014; 515: 480–482.

646. Gordon R. Give profs more money. *Winnipeg Free Press* 2008: A18, November 11.

647. Gordon R. Viewpoint: A rough analysis of funding inequity in the Faculty of Medicine, University of Manitoba. *Prairie Med J* 1996; 66: 131–132.

648. Wheeldon JP and Gordon R. Understanding research funding and folly in Canada, http://www.huffingtonpost.ca/johannes-wheeldon-phd/research-funding_b_1080238.html (2011).

649. Poulin BJ and Gordon R. *Tomorrow's Cures Today? How to Reform the Health Research System* by Donald R. Forsdyke. *Canadian Public Policy* 2000; 26: 272–273.

650. Gordon R and Poulin BJ. Quitting cold turkey: Rapid oil independence for the USA. In: Gordon R and Seckbach J (eds) *The Science of Algal Fuels: Phycology, Geology, Biophotonics, Genomics and Nanotechnology*. Dordrecht: Springer, 2012, pp. 3–20.

651. Gordon R, Poulin BJ and Wright C. World Minimum Wage (WMW). *10th Annual Meeting of the MidSouth Association of Business Disciplines (MSABD), Jackson, Mississippi.* 2004.

652. Björklund NK and Gordon R. [Nuclear state splitting: A working model for the mechanochemical coupling of differentiation "waves" with the controlling genes (master genes)] [Russian]. *Ontogenez* 1993; 24: 5–23.

653. Gordon NK and Gordon R. The organelle of differentiation in embryos: The cell state splitter [invited review]. *Theor Biol Med Model* 2016; 13. DOI: 10.1186/s12976-016-0037-2.

654. Gordon R. The ultimate Escher "Drawing Hands" problem: How does an embryo build itself? Could robots build themselves? [invited]. In: Bowermaster P (ed) *The World Transformed*. Denver: Speculist Media, 2017, https://www.google.ca/books/edition/Visions_for_a_World_Transformed/5ZnHswEACAAJ?hl=en.

655. Björklund NK and Gordon R. Surface contraction and expansion waves correlated with differentiation in axolotl embryos. I. Prolegomenon and differentiation during the plunge through the blastopore, as shown by the fate map. *Comput Chem* 1994; 18: 333–345.

656. Gordon R, Björklund NK and Nieuwkoop PD. Dialogue on embryonic induction and differentiation waves. *International Review of Cytology* 1994; 150: 373–420.

657. Nieuwkoop PD, Björklund NK and Gordon R. Surface contraction and expansion waves correlated with differentiation in axolotl embryos. II. In contrast to urodeles, the anuran *Xenopus laevis* does not show furrowing surface contraction waves. *Int J Dev Biol* 1996; 40: 661–664.

658. Gordon R, Gordon NK and Beloussov LV. Differentiation waves versus positional information gradients: Which will prevail? *BioSystems* 2018: Drafted.

659. Martin CC and Gordon R. Ultrastructural analysis of the cell state splitter in ectoderm cells differentiating to neural plate and epidermis during gastrulation in embryos of the axolotl *Ambystoma mexicanum*. *Russian Journal of Developmental Biology* 1997; 28: 71–80.

660. Gordon R. Mechanics in embryogenesis and embryonics: Prime mover or epiphenomenon? In: Podlubnaya Z (ed) *International Symposium "Biological Motility" dedicated to the memory of academician GM Frank (1904–1976), Pushchino, Moscow region, Russia, May 23 June 1, 2004*. Pushchino, Russia: Russian Academy of Sciences, 2004, pp. 190.

661. Gordon R. Mechanics in embryogenesis and embryonics: Prime mover or epiphenomenon? *Int J Dev Biol* 2006; 50: 245–253. DOI: 10.1387/ijdb.052103rg.

662. Nieuwkoop PD, Gordon R and Björklund NK. The neural induction process; its morphogenetic aspects. *Int J Dev Biol* 1999; 43: 615–623.

663. Wikipedia. Jack Butler (artist), https://en.wikipedia.org/wiki/Jack_Butler_(artist) (2024).

664. Gordon R and Butler KJ. Axolotl ectoderm contraction wave [movie], http://www.youtube.com/watch?v=XIxSXyYeOF0 (1999).

665. Martin CC and Gordon R. The evolution of perception. *Cybernetics & Systems* 2001; 32: 393–409.

666. Martin CC and Gordon R. [Ultrastructural analysis of the cell state splitter in ectoderm cells differentiating to neural plate and epidermis during gastrulation in embryos of the axolotl *Ambystoma mexicanum*] [Russian]. *Ontogenez* 1997; 28: 95–105.

667. Martin CC and Gordon R. Ultrastructural confirmation of the "cell state splitter". In: Gordon R (ed) *Symposium, Mechanics of the Cytoskeleton in Developmental Biology, Sponsored by the Canadian Society for Theoretical Biology, 34th Annual Meeting, Canadian Federation of Biological Societies, Program/Proceedings, Queen's University, Kingston, Ontario, June 9–11, 1991*. Ottawa: Canadian Federation of Biological Societies, 1991, p. 118.

668. Gordon R. Book review: Hierarchy versus Reductionism, a Review of Eldredge, N. (1995). Reinventing Darwin: The Great Debate at the High Table of Evolutionary Theory. New York: John Wiley & Sons. *J Evol Biol* 1996; 9: 383–386.

669. Noble D. *The Music of Life: Biology Beyond Genes*. Kindle ed.: OUP Oxford, 2008.

670. Gordon R and Gordon NK. The differentiation code [invited] *BioSystems* 2019; 184.

671. Alicea B and Gordon R. Cell differentiation processes as spatial networks: Identifying four-dimensional structure in embryogenesis. *BioSystems* 2018; 173: 235–246.

672. Alicea B and Gordon R. Toy models for macroevolutionary patterns and trends. *BioSystems* 2014; 122: 25–37.

673. Alicea B, McGrew S, Gordon R, Larson S, Warrington T and Watts M. DevoWorm: Differentiation waves and computation in *C. elegans* embryogenesis. *bioRxiv* 2014: https://doi.org/10.1101/009993

674. Alicea B, Gordon R and Portegys TE. Data-theoretical synthesis of the early developmental process. *Neuroinformatics* 2022; 20: 7–23.

675. Alicea B, Bastani S, Gordon NK, Crawford-Young S and Gordon R. The molecular basis of differentiation wave activity in embryogenesis. *BioSystems* 2024; 243: #10572.

676. Gordon R and Seckbach J. Biocommunication: Sign-Mediated Interactions between Cells and Organisms. London, UK: World Scientific Publishing, 2016.

677. Seckbach J and Gordon R. Preface: Who is who in biocommunication. In: Seckbach J and Gordon R (eds) *Biocommunication: Sign-Mediated Interactions between Cells and Organisms*. London: World Scientific Publishing, 2016, pp. vii-xii.

678. Gordon R. Evolution escapes rugged fitness landscapes by gene or genome doubling: The blessing of higher dimensionality. *Comput Chem* 1994; 18: 325–332.

679. Ohno S. *Evolution by Gene Duplication*. Springer Berlin Heidelberg, 2013.

680. Mihajlovic L, Iyengar BR, Baier F, Barbier I, Iwaszkiewicz J, Zoete V, Wagner A and Schaerli Y. A direct experimental test of Ohno's hypothesis. eLife Sciences Publications, Ltd, 2024.

681. Hansen TF, Carter AJR and Chiu CH. Gene conversion may aid adaptive peak shifts. *J Theor Biol* 2000; 207: 495–511. DOI: 10.1006/jtbi.2000.2189.

682. Hansen TF. Is modularity necessary for evolvability? Remarks on the relationship between pleiotropy and evolvability. *BioSystems* 2003; 69: 83–94. DOI: 10.1016/s0303-2647(02)00132-6.

683. Finkle SN, Gordon R and McAlpine PJ. Analysis of pericentric inversions. *Amer J Human Genetics* 1994; 55: A104.

684. Wikipedia. Phyllis McAlpine, https://en.wikipedia.org/wiki/Phyllis_McAlpine (2023).

685. Gordon R and Bard JBL. Establishing a HUGO Tissue Nomenclature Committee. *Human Genome News* 1999: in press, then "rejected for political reasons".

686. Gordon R. Mechanical Engineering of the Cytoskeleton in Developmental Biology. *International Review of Cytology: A Survey of Cell Biology*. San Diego, California, USA: Academic Press, 1994.

687. Gordon R. Grant agencies versus the search for truth. *Bulletin of the Canadian Society for Theoretical Biology* 1991: 4–7.

688. Wikipedia. Brian Goodwin, https://en.wikipedia.org/wiki/Brian_Goodwin (2024).

689. Goodwin B. *How the Leopard Changed Its Spots: The Evolution of Complexity*. Princeton University Press, 2001.

690. Hut P, Goodwin B and Kauffman S. Complexity and functionality: A search for the where, the when, and the how. In: BarYam Y (ed) *Unifying Themes in Complex Systems, Volume 1: Proceedings of the First International Conference on Complex Systems*. Boca Raton, FL, USA: CRC Press, 2018, pp. 259–268.

691. Goodwin BC and Brière. Mechanics of the cytoskeleton and morphogenesis of *Acetabularia*. *International Review of Cytology* 1994; 150: 225–242.

692. Beloussov LV, Saveliev SV, Naumidi II and Novoselov VV. Mechanical stresses in embryonic tissues: patterns, morphogenetic role and involvement in regulatory feedback. *Int Rev Cytol* 1994; 150: 1–34.

693. Brodland GW. Finite element methods for developmental biology. *Int Rev Cytol* 1994; 150: 95–118.

694. De Boni U. The interphase nucleus as a dynamic structure. In: Gordon R (ed) *International Review of Cytology — A Survey of Cell Biology*. 1994, pp. 149–172.

695. Ingber DE, Dike L, Hansen L, Karp S, Liley H, Maniotis A, McNamee H, Mooney D, Plopper G, Sims J and Wang N. Cellular tensegrity: Exploring how mechanical changes in the cytoskeleton regulate cell-growth, migration, and tissue pattern during morphogenesis. In: Gordon R (ed) *International Review of Cytology — A Survey of Cell Biology*. 1994, pp. 173–224.

696. Onyshko D. A possible relationship between nuclear rotation and gene activation [B.Sc. (Med) Thesis, supervisor: Richard Gordon]. *Program and Proceedings, BSc(Med) Presentations*. Winnipeg: Faculty of Medicine, University of Manitoba, 1989, pp. 142–148.

697. Maninová M, Iwanicki MP and Vomastek T. Emerging role for nuclear rotation and orientation in cell migration. *Celll Adhes Migr* 2014; 8: 42–48. DOI: 10.4161/cam.27761.

698. Preachuk C, Reed MH and Gordon R. Radiologic study of limb regeneration in the axolotl. *Axolotl Newsletter* 1994: 72–75.

699. Lyons EA and Levi CS. Ultrasound in the first trimester of pregnancy. *Radiol Clin North Am* 1982; 20: 259–270. 1982/06/01.

700. Canadian Association of Radiologists. Martin Reed, https://events. bizzabo.com/CAR14/agenda/speakers/68400 (2014).

701. Gordon R, Björklund NK, Robinson GGC and Kling HJ. PED-22125009 Y2024/M10/D11 Sheared drops and pennate diatoms. *Nova Hedwigia* 1996; 112: 287–297.

702. Manitoba Historical Society. Historic Sites of Manitoba: Mallard Lodge / Delta Marsh Field Station (RM of Portage la Prairie), https://www. mhs.mb.ca/docs/sites/deltamarshfieldstation.shtml (2024).

703. Gordon R, Björklund NK, Robinson GGC and Kling HJ. On the origin of pennate diatom shapes in shear and straining flow. In: Goldsborough LG (ed) *University of Manitoba Field Station (Delta Marsh) Twenty-Sixth Annual Report 1991*. Winnipeg: University of Manitoba, 1992, pp. 63–79.

704. Gordon R, Kling HJ and Sterrenburg FAS. A guide to the diatom literature for diatom nanotechnologists. *J Nanosci Nanotechnol* 2005; 5: 175–178.

705. Gordon R. On halting bear and professor hunts: Internet and the University of Manitoba professors' strike for academic freedom.

Negotiations (Canadian Association of University Teachers, Collective Bargaining Section) 1996; 13: 4–8.

706. Gordon R. Setting up a World Wide Site. In: Zimmerman ER and Tardif C (eds) *DRAFT Manual on Legal Strike Action for Faculty Associations*. Ottawa: Canadian Association of University Teachers, 1996, pp. 14–16.

707. Zimmerman ER. *DRAFT Manual on Legal strike Action for Faculty Associations*. Ottawa, Ontario, Canada: Canadian Association of University Teachers, 1996.

708. Abdel-Hadi M and Gordon R. Effect of 60 Hz ambient magnetic fields on the development of axolotl embryos. *Axolotl Newsletter* 1997: 10–13.

709. Abou-Ali G, Kaler KV, Paul R, Björklund NK and Gordon R. Electrorotation of axolotl embryos. *Bioelectromagnetics* 2002; 23: 214–223. DOI: 10.1002/bem.10011.

710. Levin M. Morphogenetic fields in embryogenesis, regeneration, and cancer: Non-local control of complex patterning. *BioSystems* 2012; 109: 243–261. 2012/05/01. DOI: 10.1016/j.biosystems.2012.04.005.

711. Molebny VV, Gordon R, Kurashov VN, Podanchuk DV, Kovalenko AV and Wu J. Refraction mapping of translucent objects with Shack-Harman sensor. *Proc SPIE* 1998; 3548: 31–33.

712. Murugan RM, Wexler A and Gordon R. An improved electrical impedance tomography (EIT) algorithm for the detection and diagnosis of early stages of breast cancer. In: Kinsner W (ed) *IEEE 1998 PhD Students Conference, GRADCON'98 Proceedings, May 8, 1998, Winnipeg*. IEEE, 1998.

713. Free Press Passages. Alvin Wexler, https://passages.winnipegfreepress.com/passage-details/id-301800/WEXLER_ALVIN (2021).

714. Kulbisky GP, Rickey DW, Reed MH, Björklund NK and Gordon R. The axolotl as an animal model for the comparison of 3-D ultrasound with plain film radiography. *Ultrasound Med Biol* 1999; 25: 969–975. DOI: 10.1016/s0301-5629(99)00040-x.

715. Kulbisky GP. *The axolotl salamander (Ambystoma mexicanum) as an animal model for the comparison of 3-D ultrasound with plain film radiography [B.Sc.(Med.) Thesis, Supervisors: M.H. Reed, D.W. Rickey & R. Gordon]*. Winnipeg: Faculty of Medicine, University of Manitoba, 1998.

716. Kulbisky GP, Rickey DW, Reed MH, Björklund NK and Gordon R. The axolotl as an animal model for the comparison of ultrasound with plain film radiography. *Journal of Ultrasound in Medicine* 1999; 18: S165.

717. Gordon R. The emergence of emergence: A critique of "Design, observation, surprise!". *Rivista di Biologia/Biology Forum* 2000; 93: 349–356.

718. Gordon R. Should Biologists Get ALife? A book review of: Bedau, Mark A., McCaskill, John S., Packard, Norman H., and Rasmussen, Steen (2000). Artificial Life VII: Proceedings of the Seventh International Conference on Artificial Life, Cambridge: MIT Press. 564p., US$75, pb. http://www.ALife7.ALife.org/presentations.shtml and Grand, S. (2000). Creation: Life and How to Make It, United Kingdom: Weidenfeld & Nicolson. 230p., £18.99, hb, also Harvard University Press, US$26, hb. http://www.hup.harvard.edu/catalog/GRACRE.html. *Q Rev Biol* 2002; 77: 49–50.

719. Damer B, Newman P, Gordon R, Barbalet T, Deamer DW and Norkus R. The EvoGrid: A Framework for Distributed Artificial Chemistry Cameo Simulations Supporting Computational Origins of Life Endeavors *Proceedings of the Alife XII: 12th International Conference on the Synthesis and Simulation of Living Systems, August 19–23, 2010, Odense, Denmark*. 2010, pp. 73–79.

720. Gordon R. A theory behind the bauplans of the Burgess creatures. In: Damer B (ed) *Digital Burgess, August 29-September 1, The Banff Centre for the Arts, Banff, Alberta Canada*. Banff: Banff Centre for the Arts, 1997, p. 6.

721. Ronald EMA, Sipper M and Capcarrère MS. Design, observation, surprise! A test of emergence. *Artif Life* 1999; 5: 225–239.

722. Darwin C. *On the Origin of Species by Means of Natural Selection, Or The Preservation of Favoured Races In The Struggle For Life*. 6th ed. 1872.

723. Gordon R. The fractal physics of biological evolution. In: Beysens D, Boccara N and Forgacs G (eds) *Dynamical Phenomena at Interfaces, Surfaces and Membranes*. Commack, N.Y.: NOVA Science Publishers, 1992, pp. 99–111.

724. Tomanek B, Hoult DI, Chen X and Gordon R. A probe with chest shielding for improved breast MR imaging. *Magn Reson Med* 2000; 43: 917–920.

725. Gordon R. Cosmic Embryo #2: Quitting Imported Oil Cold Turkey, http://www.science20.com/cosmic_embryo/cosmic_embryo_2_quitting_imported_oil_cold_turkey (2011, accessed August 1, 2011).

726. Gordon R, Poulin BJ and Wright C. World Minimum Wage: Ethical, economic and social considerations, illustrated by three cases. In: Arruda MC and Milton-Smith J (eds) *Third International Society of Business, Economics and Ethics (ISBEE) World Congress July 14–17, 2004, University of Melbourne, Australia*. Globethics.com, 2004.

727. Wikipedia. Books with Wings, https://en.wikipedia.org/wiki/Books_with_Wings (2023).

728. Gordon R, Hasan L, Springman M, Broughton J, Wengiel M, Ahmad B, Rawan A, Niazi W and Faqeerzai MS. Books With Wings Manual for Volunteers and Recipients, https://docs.google.com/document/d/1d-Eq1B4TfvcDF7-8-RVWNgbdXdvtUbPBwnVHSRIUHtzY/edit?hl=en_US&pli=1 (2014).

729. Gordon R and Murphy R. *Interview on Books With Wings and Higher Education for Afghans in: On Cross Country Checkup: Afghanistan. Canada plans to pull its combat troops out of Afghanistan in 2011. US secretary of state Hillary Clinton …and many Aghans want them to stay. What do you think, should Canada reconsider its role? [Audio]*. 2010.

730. Wikipedia. The Idries Shah Foundation, https://en.wikipedia.org/wiki/The_Idries_Shah_Foundation (2023).

731. Wordley P. Books With Wings UK, https://www.facebook.com/groups/BooksWithWingsCanAm/posts/748844278542347/ (2014).

732. Rawan AR, Niazi W, Sim EJ and Gordon R. Digital and book resources and training for medical libraries in Afghanistan: Digital Libraries Alliance and training through the University of Arizona and Books With Wings through the medical schools of Canada. In: Sinclair M and Bahry S (eds) *Comparative and International Education Society, 52nd Annual Conference, March 17 – 21, 2008, Gaining Educational Equity Around the World, Education in Afghanistan: The Challenges of Reform*. 2008.

733. Armstrong S. *Bitter Roots, Tender Shoots: The Uncertain Fate of Afghanistan's Women*. Toronto: Penguin Group (Canada), 2008.

734. Wikipedia. Sally Armstrong (journalist), https://en.wikipedia.org/wiki/Sally_Armstrong_(journalist) (2024).

735. Chrusch DD, Podaima BW and Gordon R. Cytobots: Intracellular robotic micromanipulators. *Canadian Conference on Electrical and Computer Engineering, 2002 IEEE CCECE 2002*. IEEE, 2002, pp. 1640–1645.

736. Podaima BW, Vaseeharan T and Gordon R. Microscopic dynamics of cytobots. *CCECE'04 Niagara Falls, ON; May 2–5, 2004*. 2004.

737. Harouche IPF, Gordon R and Shafai C. Design and simulation of a microtweezers using a controlled displacement comb drive. *CCECE/CCGEI, Ottawa, May 2006*. 2006, pp. 1071–1073.

738. Jaffe LF and Stern CD. Strong electrical currents leave the primitive streak of chick embryos. *Science* 1979; 206: 569–571.

739. Melvin C, Abdel-Hadi K, Cenzano S and Gordon R. A simulated comparison of turnstile and Poisson photons for x-ray imaging. *Canadian Conference on Electrical and Computer Engineering, 2002 IEEE CCECE 2002*. IEEE, 2002, pp. 1165–1170.

740. Lodahl P, van Driel AF, Nikolaev IS, Irman A, Overgaag K, Vanmaekelbergh DL and Vos WL. Controlling the dynamics of spontaneous emission from quantum dots by photonic crystals. *Nature* 2004; 430: 654–657. DOI: 10.1038/nature02772.

741. Oxborrow M and Sinclair AG. Single-photon sources. *Contemporary Physics* 2005; 46: 173–206. DOI: 10.1080/00107510512331337936.

742. Alkhalil FM, Perez-Barraza JI, Husain MK, Lin YP, Lambert N, Chong HMH, Tsuchiya Y, Williams DA, Ferguson AJ, Saito S and Mizuta H. Realization of Al tri-gate single electron turnstile co-integrated with a close proximity electrometer SET. *Microelectron Eng* 2013; 111: 64–67. DOI: 10.1016/j.mee.2013.02.007.

743. van Zanten DMT, Basko DM, Khaymovich IM, Pekola JP, Courtois H and Winkelmann CB. Single quantum level electron turnstile. *Physical Review Letters* 2016; 116. DOI: 10.1103/PhysRevLett.116.166801.

744. Potanina E and Flindt C. Electron waiting times of a periodically driven single-electron turnstile. *Phys Rev B* 2017; 96. DOI: 10.1103/PhysRevB.96.045420.

745. Lang L, Yingchi G, Zhichao Z, Zijun S, Chen L, Jiaqi W, Liliang G, Lan H, Chunqing G and Shiyao F. Photon total angular momentum manipulation. *Adv Photonics* 2023; 5: 056002. DOI: 10.1117/1.AP.5.5.056002.

746. Gordon R and Melvin CA. Reverse engineering the embryo: A graduate course in developmental biology for engineering students at the University of Manitoba, Canada. *Int J Dev Biol* 2003; 47: 183–187.

747. Gordon R. Conception and development of the Second Life® Embryo Physics Course. *Systems Biology in Reproductive Medicine* 2013; 59: 131–139. Review. DOI: 10.3109/19396368.2013.780644.

748. Gordon R. The fascination of diatoms: Thought experiments by an armchair diatomist [PowerPoint presentation]. In: Gaiser E (ed) *Workshop on Diatom Nanotechnology, 17th North American Diatom Symposium, October 21–26, 2003, Florida Sea Base, Lower Matecumbe Island, Florida Keys*. Miami: Florida International University, 2003.

749. Gordon R, Sterrenburg FAS and Sandhage K. A Special Issue on Diatom Nanotechnology. *J Nanosci Nanotechnol* 2005; 5: 1–4.

750. Pappas JL. A memorial to Frithjof Sterrenburg: The importance of the amateur diatomist [Chapter 1]. In: Seckbach J and Gordon R (eds) *Diatoms: Fundamentals & Applications [DIFA, Volume 1 in the series: Diatoms: Biology & Applications, series editors: Richard Gordon & Joseph Seckbach]*. Beverly, MA, USA: Wiley-Scrivener, 2019, pp. 1–27.

751. Badea C and Gordon R. Experiments with the nonlinear and chaotic behaviour of the multiplicative algebraic reconstruction technique (MART) algorithm for computed tomography. *Phys Med Biol* 2004; 49: 1455–1474. DOI: 10.1088/0031-9155/49/8/006.

752. Kojima T and Yoshinaga T. Iterative image reconstruction algorithm with parameter estimation by neural network for computed tomography. *Algorithms* 2023; 16: #60. DOI: 10.3390/a16010060.

753. Ishikawa K, Yamaguchi Y, Abou Al-Ola OM, Kojima T and Yoshinaga T. Block-iterative reconstruction from dynamically selected sparse projection views using extended power-divergence measure. *Entropy* 2022; 24: 21. DOI: 10.3390/e24050740.

754. Al-Ola OMA, Kasai R, Yamaguchi Y, Kojima T and Yoshinaga T. Image reconstruction algorithm using weighted mean of ordered-subsets EM and MART for computed tomography. *Mathematics* 2022; 10: #4277. DOI: 10.3390/math10224277.

755. Kasai R, Yamaguchi Y, Kojima T, Abou Al-Ola OM and Yoshinaga T. Noise-robust image reconstruction based on minimizing extended class of power-divergence measures. *Entropy* 2021; 23: 1005.

756. Yamaguchi Y, Kudo M, Kojima T, Abou Al-Ola O and Yoshinaga T. Extended ordered-subsets expectation-maximization algorithm with power exponent for noise-robust image reconstruction in computed tomography. 2019.

757. Kimura M, Yamaguchi Y, Abou Al-Ola OM and Yoshinaga T. Tomographic inverse problem with estimating missing projections. *Math Probl Eng* 2019; 2019. DOI: 10.1155/2019/7932318.

758. Kasai R, Yamaguchi Y, Kojima T and Yoshinaga T. Hybrid algorithm of maximum-likelihood expectation-maximization and multiplicative algebraic reconstruction technique for iterative tomographic image reconstruction. *Proc SPIE* 2019; 11049. DOI: 10.1117/12.2521185.

759. Kasai R, Yamaguchi Y, Kojima T and Yoshinaga T. Tomographic image reconstruction based on minimization of symmetrized Kullback-Leibler divergence. *Math Probl Eng* 2018; 2018: 9. DOI: 10.1155/2018/8973131.

760. Tateishi K, Yamaguchi Y, Abou Al-Ola OM, Kojima T and Yoshinaga T. Continuous analog of multiplicative algebraic reconstruction technique for computed tomography. *Conference on Medical Imaging — Physics of Medical Imaging*. San Diego, CA: Spie-Int Soc Optical Engineering, 2016.

761. Yoshinaga T, Imakura Y, Fujimoto K and Tetsushi U. Bifurcation analysis of iterative image reconstruction method for computed tomography. *Int J Bifurcation Chaos* 2008; 18: 1219–1225.

762. Yoshinaga T. Analysis of an extended PMART for CT image reconstruction as a nonlinear dynamical system. *1st International Conference on Computer Vision Theory and Applications*. Setubal, Portugal: INSTICC-Inst Syst Technologies Information Control & Communication, 2006, pp. 440–444.

763. Penner AJ, Gordon R and Kobes R. Chaos in *Bacillaria paradoxa*. In: Gaiser E (ed) *17th North American Diatom Symposium*. 2003.

764. Sivaramakrishna R. 3D breast image registration — A review. *Technol Cancer Res Treat* 2005; 4: 39–48. DOI: 10.1177/153303460500400106.

765. Sivaramakrishna R and Gordon R. The starbyte algorithm for registration of noisy breast images. *11th Canadian Conference on Electrical and Computer Engineering*. Waterloo, Canada: IEEE, 1998, pp. 673–676.

766. Sivaramakrishna R and Gordon R. Mammographic image registration using the Starbyte transformation. In: McLaren PG and Kinsner W (eds) *WESCSANEX'97*. Winnipeg, Canada: IEEE, 1997, pp. 144–149.

767. Sivaramakrishna R and Gordon R. Image registration using minimization. *IEEE WESCANEX 95 — Communications, Power, and Computing*. Winnipeg, Canada: IEEE, 1995, pp. 181–184.

768. Sivaramakrishna R and Gordon R. Mammographic image registration using a pinning function with elastic stretching. *Proceedings of the First Regional Conference, IEEE Engineering in Medicine and Biology Society, 1996 and 14th Conference of the Biomedical Engineering Society of India An International Meeting, New Delhi, India, Feb 15–18, 1995*. New York: IEEE Press, 1995, pp. 1.65–61.66.

769. Zhou X and Gordon R. Geometric unwarping for digital subtraction mammography. *Proc Vision Interface '88*. Toronto: Canadian Image Processing and Pattern Recognition Society Press, 1988, pp. 25–30.

770. Mazur AK, Mazur EJ and Gordon R. Digital differential radiography (DDR): A new diagnostic procedure for locating neoplasms, such as breast cancers, in soft, deformable tissues. *Proc SPIE* 1993; 1905: 443–455.

771. Sivaramakrishna R. Foreword: Workshop on alternatives to mammography II. *Technol Cancer Res Treat* 2005; 4: 121–122. Editorial Material.

772. Shashidhara NS, Gordon R and Sivaramakrishna R. Detection of small tumors in the breast: Combining tomography with digital subtraction. *Engineering in Medicine and Biology Society, 1996 and 14th Conference of the Biomedical Engineering Society of India An International Meeting, Proceedings of the First Regional Conference, IEEE, New Delhi, India, Feb 15–18, 1995*. New York: IEEE Press, 1995, pp. 1.51–51.52.

773. Alpuche Avilés JE, Pistorius S, Gordon R and Elbakri IA. A novel hybrid reconstruction algorithm for first generation incoherent scatter CT (ISCT) of large objects with potential medical imaging applications. *J X-Ray Sci Technol* 2011; 19: 35–56. DOI: DOI: 10.3233/XST-2010-0277.

774. Alpuche Avilés JE, Pistorius S, Elbakri IA, Gordon R and Ahmad B. A 1st generation scatter CT algorithm for electron density breast imaging which accounts for bound incoherent, coherent and multiple scatter: A Monte Carlo study. *J X-Ray Sci Technol* 2011; 19: 1–23. DOI: doi: 10.3233/XST-2011-0308.

775. Gordon R. Book review: *Breast Imaging* (A Breast Disease Book Edition). *Phys Med Biol* 2002; 47: 3565–3566. DOI: 10.1088/0031-9155/47/19/701.

776. Gordon R. Cosmic Embryo #4: Breast Cancer Wars: Delaying the TKO?, http://www.science20.com/cosmic_embryo/breast_cancer_wars_delaying_tko-78147 (2011).

777. Rangayyan RM and Gordon R. Mammographic feature enhancement. *IEEE Trans Biomed Eng* 1983; 30: 508.

778. Gordon R. Cosmic Embryo #5: Aiming for the moon with all rockets blazing? A critique of Breast Cancer Deadline 2020®, http://www.science20.com/cosmic_embryo/aiming_moon_all_rockets_blazing_critique_breast_cancer_deadline_2020-79020 (2011).

779. Gordon R and Sivaramakrishna R. Mammograms are Waldograms: Why we need 3D longitudinal breast screening [guest editorial]. *Applied Radiology* 1999; 28: 12–25.

780. Sivaramakrishna R and Shashidhar NS. Hu's moment invariants: How invariant are they under skew and perspective transformations? *1997 IEEE WESCANEX Communications, Power and Computing Conference.* Univ Manitoba, Winnipeg, Canada: IEEE, 1997, pp. 292–295.

781. Shashidhara NS. Maximizing image quality per dose in computed tomography using regularization. In: Witten M (ed) *First World Congress on Computational Medicine, Public Health and Biotechnology, April 24–28, 1994, Austin, Texas.* Austin: Center for High Performance Computing, University of Texas System, 1994.

782. Melvin C, Thulasiraman P and Gordon R. Parallel algebraic reconstruction technique for computed tomography. In: Arabnia HR and Mun Y (eds) *PDPTA'03: Proceedings of the International Conference on Parallel and Distributed Processing Techniques and Applications, Vols 1–4.* San Diego: Universal Conference Management Systems & Support, 2003, pp. 532–536.

783. Melvin C. *Design, Development and Implemenation of a Parallel Algorithm for Computed Tomography using Algebraic Reconstruction Technique [M.Sc. Thesis, Supervisors: Parimala Thulasiraman & R. Gordon]*. Winnipeg: Department of Computer Science, University of Manitoba, 2006.

784. Wikipedia. ILLIAC IV, https://en.wikipedia.org/wiki/ILLIAC_IV (2024).

785. Thulasiraman P. IDEAS (InterDisciplinary Evolving Algorithmic Science) LAB, https://home.cs.umanitoba.ca/~thulasir/ (2024).

786. Niazi WA and Gordon R. Can We Stop AIDS in Afghanistan? Presented January 14, 2005, Department of Community Health Sciences, University of Manitoba [PowerPoint], https://www.slideserve.com/laasya/can-we-stop-aids-in-afghanistan (2005).

787. Gordon R. Grasping for wholeness: A review of Stuart Pivar's book *Lifecode: The Theory of Biological Self Organization*. *Int J Dev Biol* 2006; 50: 367–368.

788. Wikipedia. Stuart Pivar, https://en.wikipedia.org/wiki/Stuart_Pivar (2024).

789. Gordon R, Bohun CS, Breward C, Cumberbatch E, Djoumna G, ElSheikh A, Williams JF and Wylie J. Yolk dynamics in amphibian embryos. *Proceedings, Fields-MITACS Industrial Problem-Solving Workshop (FMIPW), August 14–18, 2006*. Toronto: Fields Institute, 2006, pp. 33–49.

790. Gordon R. Yolk Fluid Dynamics in Amphibian Embryos. *Fields-MITACS Industrial Problem-Solving Workshop (FMIPW), Proposed Problems, August 14–18, 2006*. 2006, pp. http://www.fields.utoronto.ca/programs/scientific/06-07/FMIPW/proposed_problems/index.html#03.

791. Gordon R, Luppes R, Veldman AEP, Tuszynski JA and Nouri M. Space travel and the origin of bilateral symmetry: yolk fluid dynamics in amphibian embryos calculated by a two-phase finite element simulation. In: Gumel A (ed) *Fields Institute & MITACS Workshop on Industrial Mathematics, August 14–18, 2006, Toronto*. 2006.

792. Bohun CS, Breward C, Cumberbatch E, Djoumna G, ElSheikh A, Gordon R, Williams JF and Wylie J. Yolk Dynamics in Amphibian Embryos [PowerPoint Presentation]. *Presentation, Fields-MITACS*

Industrial Problem-Solving Workshop (FMIPW), August 14–18, 2006. Toronto: Fields Institute, 2006.

793. Bohun CS and Breward CJW. Yolk dynamics in amphibian embryos. *Mathematics-in-Industry Case Studies Journal* 2009; 1: 99–119.

794. Nouri C, Tuszynski JA, Weibe M and Gordon R. Simulation of the role of microtubules in the cortical rotation of amphibian embryos in normal and zero gravity. *Computational Cell Biology: Abstracts of Papers Presented at the 2009 Meeting Held March 24–27, 2009.* Cold Spring Harbor, NY: Cold Spring Harbor Laboratory, 2009.

795. Björklund NK and Gordon R. A hypothesis linking low folate intake to neural tube defects due to failure of post-translation methylations of the cytoskeleton. *Int J Dev Biol* 2006; 50: 135–141.

796. Gisondi P, Fantuzzi F, Malerba M and Girolomoni G. Folic acid in general medicine and dermatology. *J Dermatol Treat* 2007; 18: 138–146. DOI: doi:10.1080/09546630701247930.

797. Yakoob MY, Menezes EV, Soomro T, Haws RA, Darmstadt GL and Bhutta ZA. Reducing stillbirths: behavioural and nutritional interventions before and during pregnancy. *BMC Pregnancy Childbirth* 2009; 9 Suppl 1: S3. DOI: 10.1186/1471-2393-9-S1-S3.

798. Mayanil CS, Ichi S, Farnell BM, Boshnjaku V, Tomita T and McLone DG. Maternal intake of folic acid and neural crest stem cells. In: Litwack G (ed) *Vitamins and Hormones: Stem Cell Regulators.* San Diego: Elsevier Academic Press Inc, 2011, pp. 143–173.

799. Imbard A, Benoist JF and Blom HJ. Neural tube defects, folic acid and methylation. *Int J Environ Res Public Health* 2013; 10: 4352–4389. DOI: 10.3390/ijerph10094352.

800. Strickland KC, Krupenko NI and Krupenko SA. Molecular mechanisms underlying the potentially adverse effects of folate. *Clin Chem Lab Med* 2013; 51: 607–616. DOI: 10.1515/cclm-2012-0561.

801. Cao R, Xie J and Zhang L. Abnormal methylation caused by folic acid deficiency in neural tube defects. *Open Life Sci* 2022; 17: 1679–1688. DOI: 10.1515/biol-2022-0504.

802. Rai S, Leydier L, Sharma S, Katwala J and Sahu A. A quest for genetic causes underlying signaling pathways associated with neural tube defects. *Front Pediatr* 2023; 11. DOI: 10.3389/fped.2023.1126209.

803. Tromp JT and Gordon R. The number of 3D configurations of a labeled size 2*n "Wurfel" [A117613: The On-Line Encyclopedia of Integer Sequences], http://oeis.org/A117613 (2006).

804. Tromp J. John Tromp, https://tromp.github.io/ (2024).

805. Van Uytven E, Pistorius S and Gordon R. An iterative three-dimensional electron density imaging algorithm using uncollimated Compton scattered x rays from a polyenergetic primary pencil beam. *Med Phys* 2007; 34: 256–274.

806. Van Uytven E, Pistorius S and Gordon R. Prediction of multiple scatter distributions using an iterative Monte Carlo technique for Compton scatter mammography. *Med Phys* 2006; 33: 2670.

807. Van Uytven E. *3D Electron Density Imaging using Single Scattered X rays with Application to Breast CT and Mammographic Screening [Ph.D. Thesis, Supervisors: S. Pistorius & R. Gordon]*. Winnipeg: Department of Physics and Astronomy, University of Manitoba, 2007.

808. Van Uytven E, Pistorius S and Gordon R. A method for 3D electron density imaging using single scattered X rays with application to mammographic screening. *Phys Med Biol* 2008; 53: 5445–5459.

809. Gordon R, Hoover RB, Tuszynski JA, de Luis J, Camp PJ, Tiffany MA, Nagy SS, Fayek M, Lopez PJ and Lerner BE. Diatoms in space: Testing prospects for reliable diatom nanotechnology in microgravity. *Proceedings of the SPIE* 2007; 6694: V1-V15, doi:10.1117/1112.737051.

810. Kimmich R. Diatom Forum, https://groups.io/g/diatom-forum/plans (2024).

811. Wikipedia. File:Diatoms through the microscope.jpg, https://commons.wikimedia.org/wiki/File:Diatoms_through_the_microscope.jpg (2024).

812. Parkinson J and Gordon R. Beyond micromachining: The potential of diatoms. *Trends Biotechnol (Tibtech)* 1999; 17: 190–196.

813. Gordon R and Parkinson J. Potential roles for diatomists in nanotechnology. *J Nanosci Nanotechnol* 2005; 5: 35–40.

814. Vinayak V, Joshi KB, Gordon R and Schoefs B. Nanoengineering the diatom surface for plausible applications. In: Losic D (ed) *Diatom Nanotechnology: Progress and Emerging Applications*. London: Royal Society of Chemistry, 2018, pp. 55–78.

815. Gordon R, Crawford-Young S and Nouri C. *Can we have babies in space? Embryogenesis in microgravity/Embryos and Cells in Microgravity/ Rayleigh-Taylor Instability of an Inverted Axolotl Egg, Pediatric Research Rounds, June 14, 2007 [PowerPoint]*. Winnipeg: Manitoba Institute for Child Research, 2007.

816. Crawford-Young S. Effects of microgravity on cell cytoskeleton and embryogenesis. *Int J Dev Biol* 2006; 50: 183–191.

817. Gordon R and Hoover RB. Could there have been a single origin of life in a Big Bang universe? *Proceedings of the SPIE* 2007; 6694. DOI: doi:10.1117/12.737041.

818. Wainwright M, Rose CE, Baker AJ and Wickramasinghe NC. Biological entities isolated from the stratosphere (22–27km) -Case for their space origin. *Proc SPIE* 2013; 8865. DOI: 10.1117/12.2047304.

819. Gordon R. Critique of a claimed discovery of a diatom from outer space, http://www.kurzweilai.net/critique-of-a-claimed-discovery-of-a-diatom-from-outer-space (2013).

820. Gordon R and Tuszynski JA. Preface 3: Monotheism: The basis for unifying Abrahamic religion and science? In: Seckbach J and Gordon R (eds) *Divine Action and Natural Selection: Science, Faith and Evolution*. Singapore: World Scientific, 2008, p. lii.

821. Seckbach J and Gordon R. Divine Action and Natural Selection: Science, Faith and Evolution [DINA]. Singapore: World Scientific, 2008.

822. Tuszynski JA, Gordon R and Berger D. Dialogue on: "A Cabbalistic interpretation of the theory of evolution based on the teachings of Rabbi Kook: natural selection and its moral implication" by Dov Berger. In: Seckbach J and Gordon R (eds) *Divine Action and Natural Selection: Science, Faith and Evolution*. Singapore: World Scientific, 2008, pp. 203–215.

823. Gordon R. Preface 2: To the scientist who feels above the creationist debate. In: Seckbach J and Gordon R (eds) *Divine Action and Natural Selection: Science, Faith and Evolution*. Singapore: World Scientific, 2008, pp. xxxviii-xliii.

824. Gordon R. Over-confident anti-creationists versus over-confident creationists [Chapter 12]. In: Seckbach J and Gordon R (eds) *Divine Action and Natural Selection: Science, Faith and Evolution*. Singapore: World Scientific, 2008, pp. 216–247.

825. Seckbach J and Gordon R. Theology and Science: From Genesis To Astrobiology [TASE]. In: Seckbach J and Gordon R, (eds.). Singapore: World Scientific Publishing, 2019.

826. Swan L, Gordon R and Seckbach J. Origin(s) of Design in Nature: A Fresh, Interdisciplinary Look at How Design Emerges in Complex Systems, Especially Life [ODIN]. Dordrecht, Netherlands: Springer, 2012.

827. Seckbach J. Origins: Genesis, Evolution and Diversity of Life (Cellular Origin, Life in Extreme Habitats and Astrobiology). Springer, 2004.

828. Sterrenburg F, Gordon R, Tiffany MA and Nagy SS. Diatoms: Living in a constructal environment. In: Seckbach J (ed) *Algae and Cyanobacteria in Extreme Environments*. Springer, 2007, pp. 141–172.

829. Fleury V and Gordon R. Coupling of growth, differentiation and morphogenesis: An integrated approach to design in embryogenesis. In: Swan L, Gordon R and Seckbach J (eds) *Origin(s) of Design in Nature: A Fresh, Interdisciplinary Look at How Design Emerges in Complex Systems, Especially Life*. Dordrecht: Springer, 2012, pp. 385–428.

830. Nouri C, Luppes R, Veldman AEP, Tuszynski JA and Gordon R. Rayleigh instability of the inverted one-cell amphibian embryo. *Phys Biol* 2008; 5: #015006. DOI: 10.1088/1478-3975/5/1/015006.

831. Gordon R, Luppes R, Veldman AEP and Tuszynski JA. Origin of bilateral symmetry: Yolk fluid dynamics in amphibian embryos calculated by a two-phase finite element simulation. In: Beloussov LV and Gordon R (eds) *International Conference on the Physical Foundations of Development and Cell Differentiation, International Institute of Biophysics (IIB) and the Russian Academy of Natural Sciences, Neuss, Germany, August 19–23, 2006*. 2006.

832. Luppes R. dr. ir. R. (Roel) Luppes, https://www.rug.nl/staff/r.luppes/?lang=en (2024).

833. Nouri C, Tuszynski JA, Wiebe M and Gordon R. Simulation of the effects of microtubules in the cortical rotation of amphibian embryos in normal and zero gravity. *BioSystems* 2012; 109: 444–449. 2012/06/09. DOI: 10.1016/j.biosystems.2012.05.009.

834. Tuszynski JA and Gordon R. A mean field Ising model for cortical rotation in amphibian one-cell stage embryos. *BioSystems* 2012; 109: 381–389.

835. Tuszynski JA. The Emerging Physics of Consciousness. Heidelberg: Springer Verlag, 2006.

836. Gordon R. Consciousness of the one cell axolotl embryo [PowerPoint presentation]. *The 2009 Hackers Conference — The 25th Conference, Santa Cruz, California, November 6–8, 2009.* 2009.

837. Falk D. Insects and other animals have consciousness, experts declare, https://www.quantamagazine.org/insects-and-other-animals-have-consciousness-experts-declare-20240419/?mc_cid=7312d47f39 (2024).

838. Stanford University. Comron Nouri, https://explorecourses.stanford.edu/instructor/comron (2024).

839. Neethirajan S, Gordon R and Wang L. Potential of silica bodies (phytoliths) for nanotechnology. *Trends Biotechnol* 2009; 27: 461–467. DOI: doi:10.1016/j.tibtech.2009.05.002.

840. Gordon R and Westfall JE. Google Embryo for building quantitative understanding of an embryo as it builds itself: I. Lessons from Ganymede and Google Earth. *Biological Theory: Integrating Development, Evolution, and Cognition* 2009; 4: 390–395 + Supplementary Appendix.

841. Gordon R. Whole embryo surface. *Photon Spect* 1996; 30: 73.

842. Gordon R. Photonics Problem-Solver: Whole embryo surface. *Photon Spect* 1996; 30: 73.

843. Gordon R. Google Embryo for building quantitative understanding of an embryo as it builds itself: II. Progress toward an embryo surface microscope. *Biological Theory: Integrating Development, Evolution, and Cognition* 2009; 4: 396–412.

844. Crawford-Young S. *A Robotic Microscope for 3D Time-Lapse Imaging of Early Stage Axolotl Salamander Embryos [M.Sc. Thesis, Supervisor: R. Gordon]*. Winnipeg: Department of Electrical & Computer Engineering, University of Manitoba, 2007.

845. Crawford-Young SJ, Weatherbee A, Fleuraru C, Gordon R, Thanusutiyabhorn P, Vitkin A, Mao Y and Sherif SS. Non-destructive imaging of live early-stage axolotl salamander embryo using Optical Coherence Tomography [Poster 255-TCSB-73, Paper Code: 40.11].

Proceedings, 18th Photonics North Conference, Québec City, May 24–26, 2016. PhotonicsNorth, 2016, https://www.conferium.com/uploads/DocsConv_255/255-TCSB-273.doc.

846. Ramachandra TV, Mahapatra DM, Karthick B. and Gordon R. Milking diatoms for sustainable energy: Biochemical engineering versus gasoline-secreting diatom solar panels. *Ind Eng Chem Res* 2009; 48: 8769–8788.

847. Green Car Congress. Researchers Propose Milking Diatoms to Yield Massive Amounts of Oil or Bio-Hydrocarbon Fuels, http://www.greencarcongress.com/2009/06/diatoms-20090618.html (2009).

848. Gordon R. Cost not the problem. *Winnipeg Free Press*, May 11, 2009: A11.

849. Gordon R, Merz CR, Gurke S and Schoefs B. Bubble farming: Scalable microcosms for diatom biofuel and the next Green Revolution [Chapter 22]. In: Seckbach J and Gordon R (eds) *Diatoms: Fundamentals & Applications [DIFA, Volume 1 in the series: Diatoms: Biology & Applications, series editors: Richard Gordon & Joseph Seckbach].* Beverly, MA, USA: Wiley-Scrivener, 2019, pp. 583–654.

850. Vinayak V, Gordon R, Schoefs B and Joshi KB. Diafuel: Trade Mark No. 3778882 in respect of Biofuels, (2018).

851. Vinayak V, Manoylov KM, Gateau H, Blanckaert V, Hérault J, Pencréac'h G, Marchand J, Gordon R and Schoefs B. Correction: Vinayak, V., *et al.* Diatom Milking: A Review and New Approaches. Marine Drugs 2015, 13, 2629–2665. *Mar Drugs* 2015; 13: 7301.

852. Khan MJ, Gordon R, Varjan S and Vinayak V. Employing newly developed plastic bubble wrap technique for biofuel production from diatoms cultivated in discarded plastic waste bubble wraps collected from bubble wrap industry. *Sci Total Environ* 2022; 823.

853. Vinayak V, Kumar V, Kashyap M, Joshi KB, Gordon R and Schoefs B. Fabrication of resonating microfluidic chamber for biofuel production in diatoms (Resonating device for biofuel production). In: Ganguly S and Saha D (eds) *3rd International Conference on Emerging Electronics (ICEE), December 27–30, 2016, IIT Bombay, Mumbai, India.* IEEE Electron Devices Society, 2017, pp. doi: 10.1109/ICEmElec.2016.8074628

854. Vinayak V, Manoylov KM, Gateau H, Blanckaert V, Pencréac'h G, Hérault J, Marchand J, Gordon R and Schoefs B. Diatom milking: A

review and new approaches [Correction: (2015), 13(12), 7301]. *Mar Drugs* 2015; 13: 2629–2665. DOI: 10.3390/md13052629.

855. Vinayak V, Gordon R, Gautam S and Rai A. Discovery of a diatom that oozes oil. *Adv Sci Lett* 2014; 20: 1256–1267. DOI: 10.1166/asl.2014.5591.

856. Vinayak V, Mishra V, Gautam S, Rai A and Gordon R. Oozing of oil from diatoms without cell lysis. In: Khan ZH, Husain M, Siddiqui WA, *et al.* (eds) *Advances in Nanotechnology and Renewable Energy, Conference Proceedings (NCNRE-14, National Conference on Nanotechnology and Renewable Energy) (April 28–29, 2014)*. Delhi: Bharti Publications, 2014, pp. 628–634.

857. Vinayak V and Gordon R. Diatom solar panels. In: Yadava RN (ed) *National symposium on Horizons of light in molecules, materials and daily life, December 18–19, 2015, Department of Chemistry (School of Chemical Sciences & Technology), Dr H S Gour Central University, Sagar, MP 470003, India*. Sagar, MP India: Department of Chemistry (School of Chemical Sciences & Technology), Dr. H. S. Gour Central University, 2015.

858. Merz C, Arora N, Welch M, Lo E and Philippidis G. Microalgal cultivation characteristics using commercially available air-cushion packaging material as a photobioreactor. *Sci Rep* 2023; 13. DOI: 10.1038/s41598-023-30080-6.

859. Tiffany MA, Gordon R and Gebeshuber IC. *Hyalodiscopsis plana*, a sublittoral centric marine diatom, and its potential for nanotechnology as a natural zipper-like nanoclasp [correction: Fig. 22 scale bar = 5 µm]. *Polish Botanical Journal* 2010; 55: 27–41.

860. Gebeshuber I, Zischka F, Tiffany MA and Ghobara M. Colonial Diatoms [DCHN, Volume in the series: Diatoms: Biology & Applications, series editors: Richard Gordon & Joseph Seckbach, In preparation]. In: Gordon R and Seckbach J, (eds.). *Diatoms: Biology & Applications*. Beverly, MA, USA: Wiley-Scrivener, 2025.

861. Wikipedia. Ille Gebeshuber, https://en.wikipedia.org/wiki/Ille_Gebeshuber (2023).

862. Futterknecht O, Anderson P, Gordon R and Gebeshuber IC. Teeth and injectors across scales: Inspiration from biology for emerging grinders, cutters and drills with optimized tribological performance. *5th*

Vienna International Conference on Nanotechnology, Viennano '13, Palaoplimpico, Torino, Italy, September 8–13, 2013. 2013, pp. submitted, rejected as "out of scope".

863. Gebeshuber IC and Gordon R. Bioinspiration for tribological systems on the micro- and nanoscale: Dynamic, mechanic, surface and structure related functions. *Micro and Nanosystems* 2011; 3: 271–276.

864. Gordon R, Witkowski A, Gebeshuber IC and Allen CS. The diatoms of Antarctica and their potential roles in nanotechnology. In: Masó M and Chillida A (eds) *Antarctica: Time of Change*. Barcelona: Editions ACTAR, 2010, pp. 84–95.

865. Gebeshuber IC, Gordon R and Macqueen MO. Algae Biofuels. In: Mulvaney D and Robbins P (eds) *Green Technology: An A-to-Z Guide, eReference, SAGE Series on Green Society Vol 10*. Thousand Oaks, California: SAGE eReference, Sage Publications, 2011, http://www.sagepub.com/booksProdDesc.nav?prodId=Book234026.

866. Gordon R, Levin SM and Gebeshuber IC. Triggered, nanostructured biodegradables (TNBs) for surgical implants. *Micro and Nanosystems* 2011; 3: 284–289.

867. Zischka F, Kratochvil H, Noll A, Gordon R, Harbich T and Gebeshuber IC. Diatom triboacoustics [Chapter 11]. In: Cohn SA, Manoylov KM and Gordon R (eds) *Diatom Gliding Motility [DIGM, Volume in the series: Diatoms: Biology & Applications, series editors: Richard Gordon & Joseph Seckbach] Wiley-Scrivener, Beverly, MA, USA*. Beverly, Massachusetts, USA: Wiley-Scrivener, 2021, pp. 249–282.

868. Carron I and Sidky E. CS: ART and CS in CT, Regularization for Matrix Completion, Estimation of High-Dimensional Low-Rank Matrices, http://nuit-blanche.blogspot.com/2010/01/cs-art-and-cs-in-ct-regularization-for.html (2010).

869. Carron I and Gordon R. MIA'09 and Richard Gordon's ART CT algorithm, http://nuit-blanche.blogspot.ca/2009/12/mia09-and-richard-gordons-art-ct.html (2009).

870. Gordon R. Comment on "CS: These Technologies Do Not Exist: Random X-ray Collectors in CT", http://nuit-blanche.blogspot.com/2009/11/cs-these-technologies-do-not-exist_28.html (2009).

871. Carron I and Gordon R. Breaking the coherence barrier: asymptotic incoherence and asymptotic sparsity in compressed sensing, http://nuit-blanche.blogspot.com/2013/02/breaking-coherence-barrier-asymptotic.html (2013).

872. Carron I, Gordon R and Raskar R. CS: Convergence Analysis of SART by Bregman Iteration and Dual Gradient Descent, http://nuit-blanche.blogspot.com/2010/06/cs-convergence-analysis-of-sart-by.html (2010).

873. Carron I and Gordon R. CS: These Technologies Do Not Exist: Random X-ray Collectors in CT, http://nuit-blanche.blogspot.com/2009/11/cs-these-technologies-do-not-exist_28.html (2009, accessed January 26, 2011).

874. Koetzier L, Mastrodicasa D, Szczykutowicz T, Werf N, Wang A, Sandfort V, Molen A, Fleischmann D and Willemink M. Deep learning image reconstruction for CT: Technical principles and clinical prospects. *Radiology* 2023: doi: 10.1148/radiol.221257. DOI: 10.1148/radiol.221257.

875. Wikipedia. John von Neumann, https://en.wikipedia.org/wiki/John_von_Neumann (2024).

876. CancerCareManitoba. Jorge E. Alpuche Aviles, https://www.cancercare.mb.ca/For-Health-Professionals/radiation-oncology-program-team/medical-physicists-bios (2024).

877. Gordon R. Epilogue: The diseased breast lobe in the context of X-chromosome inactivation and differentiation waves. In: Tot T (ed) *Breast Cancer: A Lobar Disease*. London: Springer, 2011, pp. 205–210.

878. Lu K, Cao T and Gordon R. A cell state splitter and differentiation wave working-model for embryonic stem cell development and somatic cell epigenetic reprogramming. *BioSystems* 2012; 109: 390–396.

879. Lu K, Gordon R and Cao T. Reverse engineering the mechanical and molecular pathways in stem cell morphogenesis. *J Tissue Eng Regen Med* 2013: doi: 10.1002/term.1672.

880. Lu K, Gordon R and Cao T. Reverse engineering the mechanical and molecular pathways in stem cell morphogenesis. *J Tissue Eng Regen Med* 2015; 9: 169–173. DOI: 10.1002/term.1672.

881. Gordon R. Cosmic Embryo #1: My Erdös Number Is 2*i*, http://www.science20.com/cosmic_embryo/cosmic_embryo_1_my_erdös_number_2i-75020 (2011).

882. NASA and ESA. V838 MONOCEROTIS revisited: Space phenomenon imitates art, https://sci.esa.int/web/hubble/-/34789-v838-monocerotis-revisited (2019).

883. Goffman C. And what is your Erdös number? *Am Math Mon* 1969; 76: 791.

884. Wikipedia. Stanisław Marcin Ulam, https://en.wikipedia.org/wiki/Stanis%C5%82aw_Ulam (2024).

885. jewornotjew.com. Stanislaw Ulam, http://www.jewornotjew.com/profile.jsp?ID=2303 (2016).

886. Kornberg H. Memoirs of a biochemical Hod carrier. *The Journal of biological chemistry* 2003; 278: 9993–10001. DOI: 10.1074/jbc.X200008200.

887. Regents of the University of Colorado. Jack Sadler Memorial Lecture Series, https://medschool.cuanschutz.edu/biochemistry/upcoming-events/jack-sadler-memorial-lecture-series (2024).

888. Gordon R. Cosmic Embryo #3: The ART of 3D Sun and Breast Cancer Imaging, http://www.science20.com/cosmic_embryo/cosmic_embryo_3_art_3d_sun_and_breast_cancer_imaging-75977 (2011).

889. Wikipedia. Joshua Lederberg, https://en.wikipedia.org/wiki/Joshua_Lederberg (2023).

890. Gordon R. Diatoms and nanotechnology: Early history and imagined future as seen through patents. In: Smol JP and Stoermer EF (eds) *The Diatoms: Applications for the Environmental and Earth Sciences*. 2nd ed. Cambridge: Cambridge University Press, 2010, pp. 585–602.

891. Pappas JL. *Diatom Arts and Crafts [DART, Volume in the series: Diatoms: Biology & Applications, series editors: Richard Gordon & Joseph Seckbach, In preparation]*. Beverly, MA, USA: Wiley-Scrivener, 2025.

892. Gordon R, Rudloe J and Carron I. Octopus art via compressive sensing, http://nuit-blanche.blogspot.ca/2012/03/nuit-blanche-readers-mailbag.html (2012, accessed March 14, 2012).

893. Carron I and Gordon R. Compressive Sensing Literature This Week: The application of compressive sampling to radio astronomy I: Deconvolution, http://www.blogger.com/comment.g?blogID=6141980&postID=303990642450925592&page=1&token=1308037879179 (2011).

894. Amazon.com. Jack Rudloe's books, https://www.amazon.com/s?k=jack+rudloe&dc&crid=32H9HZSZUOYT8&sprefix=jack+rudloe%2Caps%2C138&ref=a9_asc_1 https://www.amazon.com/s?k=jack+rudloe&dc&page=2&crid=32H9HZSZUOYT8&qid=1728691779&sprefix=jack+rudloe%2Caps%2C138&ref=sr_pg_1 (2024).

895. Manning KR. *Black Apollo of Science. The Life of Ernest Everett Just*. New York: Oxford University Press, 1983.

896. Stazione Zoologica Anton Dohrn Napoli — Italy. Luigia Santella, https://www.szn.it/index.php/en/staff/list-of-personnel/359-santella-luigia/441-santella-luigia (2015).

897. Wikipedia. Bryostatin, https://en.wikipedia.org/wiki/Bryozoa (2024).

898. Ruechel J. *Plunderers of the Earth: the Erosion of Civilization, the Mad Crusade to Control the Climate, and the Untold Story of CO_2*. Kindle Edition 2024.

899. Gulf Specimen Marine Lab. Gulf Specimen Marine Lab Staff, https://gulfspecimen.org/staff/ (2024).

900. Gordon R. The Living Jack: A Celebration of Jack Rudloe at 70 and His Living Dock. Panacea, Florida, USA: Gulf Specimen Marine Laboratory, 2014.

901. Rudloe J. *The Living Dock at Panacea*. Knopf, 1977.

902. Rudloe J and Anderson W. *The Living Dock*. Great Outdoors Publishing Company, 2003.

903. Wikipedia. Jack Rudloe, https://en.wikipedia.org/wiki/Jack_Rudloe (2023).

904. Gordon R. The vanishing physician scientist: A critical review and analysis. *Accountability in Research: Policies and Quality Assurance* 2012; 19: 89–113. DOI: DOI: 10.1080/08989621.2012.660076.

905. Schafer AI. The Vanishing Physician-Scientist? Ithaca, NY, USA: Cornell University Press, 2009.

906. Zhang WY, Jiang Y and Chen JZY. Onsager model for the structure of rigid rods confined on a spherical surface. *Physical Review Letters* 2012; 108: 057801. DOI: 10.1103/PhysRevLett.108.057801.

907. Gordon R. Conception and development of the Second Life® Embryo Physics Course [invited]. *Systems Biology in Reproductive Medicine* 2013; 59: 131–139.

908. Rudnyi EB. Embryogenesis Explained, http://embryogenesisexplained. rudnyi.ru/old-lectures/ (2023).

909. Gordon R. Cosmic Embryo #6: Swaying Trees to Produce Wind Energy, http://www.science20.com/cosmic_embryo/cosmic_embryo_6_sway-ing_trees_produce_wind_energy-107333 (2013).

910. McGarry S and Knight C. The potential for harvesting energy from the movement of trees. *Sensors* 2011; 11: 9275–9299. DOI: 10.3390/s111009275.

911. Ghobara M, Gordon R and Reissig L. Diatom frustules: A biomaterial with promising photonic properties [poster]. In: Matijević M, Krstić M and Beličev P (eds) *Book of abstracts, PHOTONICA2019 , The Seventh International School and Conference on Photonics, 26 August – 30 August 2019, Belgrade, Serbia & Machine Learning with Photonics Symposium (ML-Photonica 2019)* Belgrade, Serbia: Vinča Institute of Nuclear Sciences, 2019, pp. 118.

912. Gordon R. Cosmic Embryo #7: Is there an optimum solution to the missing sock problem?, http://www.science20.com/cosmic_embryo/blog/cosmic_embryo_7_is_there_an_optimum_solution_to_the_miss-ing_sock_problem-133100 (2014).

913. Gordon R. Soap bar skin scanner for detection of early melanoma. *Journal of Brief Ideas* 2016: https://zenodo.org/records/154549.

914. Seckbach J and Gordon R. Introduction to biocommunication. In: Seckbach J and Gordon R (eds) *Biocommunication: Sign-Mediated Interactions between Cells and Organisms*. London: World Scientific Publishing, 2016, pp. xiii-xvi.

915. Gordon R and Stone R. Cybernetic embryo [Chapter 5]. In: Gordon R and Seckbach J (eds) *Biocommunication: Sign-Mediated Interactions between Cells and Organisms*. London: World Scientific Publishing, 2016, pp. 111–164.

916. Koestler A. *Janus: A Summing Up*. New York, NY, USA: Random House, 1978.

917. Portegys T, Pascualy G, Gordon R, McGrew S and Alicea B. Morphozoic, cellular automata with nested neighborhoods as a metamorphic repre-sentation of morphogenesis [invited]. In: Adamatti DF (ed) *Multi-Agent*

Based Simulations Applied to Biological and Environmental Systems. IGI Global, 2016, pp. 44–80.

918. Gutiérrez A, Gordon R and Dávila LP. Deformation modes and structural response of diatom shells. *Journal of Materials Science and Engineering with Advanced Technology* 2017; 15: 105–134.

919. Alvare G and Gordon R. Foxels for high flux, high resolution computed tomography (FoxelCT) using broad x-ray focal spots: Theory and two-dimensional fan beam examples. *Radiology and Diagnostic Imaging* 2017; 1: https://www.oatext.com/foxels-for-high-flux-high-resolution-computed-tomography-foxelct-using-broad-x-ray-focal-spots.php. DOI: doi: 10.15761/RDI.1000103.

920. Mazur EJ and Gordon R. Interpolative algebraic reconstruction techniques without beam partitioning for computed tomography. *Med Biol Eng Comput* 1995; 33: 82–86.

921. Wang C and Gordon R. RICT: Rotating image computed tomography with a one-to-one reversible image rotation algorithm. *J X-Ray Sci Technol* 2023; 31: 463–482. DOI: 10.3233/XST-221248.

922. Düchting H. *Georges Seurat, 1859–1891: The Master of Pointillism.* Köln, Germany: Taschen, 2001.

923. Luchka KB. *Application of Digital Image Processing Techniques in Clinical Radiotherapy [M.Sc. Thesis, Supervisors: Shlomo Shalev, Richard Gordon].* Winnipeg: Departments of Physics and Medical Physics, University of Manitoba, 1996.

924. Rostamzadeh M, Luchka K, Ma R, Liu M, Dunne E, Camborde ML, Karan T, Mestrovic A and Bergman A. In-vivo quality assurance of dynamic tumor tracking (DTT) for liver SABR using EPID images. *J Appl Clin Med Phys* 2023; 24: #e13969. DOI: 10.1002/acm2.13969.

925. Igamberdiev AU, Gordon R, Alicea B and Cherdantsev VG. Computational, theoretical, and experimental approaches to morphogenesis. *BioSystems* 2018; 173: 1–3. DOI: 10.1016/j.biosystems.2018.09.018.

926. Igamberdiev AU, Beloussov LV and Gordon R. Special Issue: Biological Morphogenesis: Theory and Computation. *BioSystems* 2012; 109: 241–242. DOI: 10.1016/j.biosystems.2012.07.002.

927. Ricercatore. File:Igamberdiev2011 Wartburg1.jpg, https://commons.wikimedia.org/wiki/File:Igamberdiev2011_Wartburg1.jpg (2019).

928. Sharov AA. Genome increase as a clock for the origin and evolution of life. *Biol Direct* 2006; 1. DOI: 10.1186/1745-6150-1-17.

929. Gordon R. Book proposal: Habitability of the Universe Before Earth [poster]. In: Walker S (ed) *Re-conceptualizing the Origin of Life, November 9–13, 2015*. Washington, DC, USA: Carnegie Institution for Science, 2015.

930. Sharov A and Mikhailovsky G. Pathways to the Origin and Evolution of Meanings in the Universe [Volume in the series: Astrobiology Perspectives on Life of the Universe book series (Series editors: Richard Gordon & Joseph Seckbach, Wiley-Scrivener)]. Beverly, Massachusetts, USA: Wiley-Scrivener, 2024.

931. Sharov AA. Chapter 17: Coenzyme world model of the origin of life [Reprint of: Sharov, A.A., 2016. Coenzyme world model of the origin of life. BioSystems 144, 8–17.]. In: Gordon R and Sharov AA (eds) *Habitability of the Universe Before Earth [in series: Astrobiology: Exploring Life on Earth and Beyond, eds Pabulo Henrique Rampelotto, Joseph Seckbach & Richard Gordon]*. Amsterdam: Elsevier B.V., 2018, pp. 407–426.

932. Mikhailovsky GE and Gordon R. Symbiosis: Why was the transition from microbial prokaryotes to eukaryotic organisms a cosmic gigayear event? In: Gordon R and Sharov AA (eds) *Habitability of the Universe Before Earth [in series: Astrobiology: Exploring Life on Earth and Beyond, eds Pabulo Henrique Rampelotto, Joseph Seckbach & Richard Gordon]*. Amsterdam, Netherlands: Elsevier B.V., 2018, pp. 355–405.

933. Mikhailovsky GE and Gordon R. LUCA to LECA, the Lucacene: A model for the gigayear delay from the first prokaryote to eukaryogenesis. *BioSystems* 2020; 205.

934. Mikhailovsky G and Gordon R. Shuffling type of biological evolution based on horizontal gene transfer and the biosphere gene pool hypothesis. *BioSystems* 2020: #104131.

935. Mikhailovsky GE, Gordon R and Igamberdiev AU. Editorial: Symbiogenesis and progressive evolution. *BioSystems* 2021, #104429.

936. Gordon R. Cosmic Embryo #8: What's Swimming On/In My Eyes?, https://www.science20.com/richard_dick_gordon/cosmic_embryo_8_whats_swimming_onin_my_eyes-234966 (2018).

937. Annenkov VV, Danilovtseva EN and Gordon R. Steps of silicic acid transformation to silica frustules: Main hypothesis and discoveries. In: Annenkov V, Seckbach J and Gordon R (eds) *Diatom Morphogenesis [DIMO, Volume in the series: Diatoms: Biology & Applications, series editors: Richard Gordon & Joseph Seckbach]*. Beverly, MA, USA: Wiley-Scrivener, 2020, pp. 301–347.

938. Annenkov VV, Gordon R, Zelinskiy SN and Danilovtseva EN. The probable mechanism for silicon capture by diatom algae: Assimilation of polycarbonic acids with diatoms, is endocytosis a key stage in building of siliceous frustules? . *J Phycol* 2020; 56: 1729–1737. DOI: https://doi.org/10.1111/jpy.13062.

939. Harbich T, Gordon R, Cohn SA, Ashworth MP, Annenkov V and Goessling JW. "Diatoms: Life in Glass Houses" revisited: Updates and comments. *Limnology and Freshwater Biology* 2024, 1, 1454–1470.

940. Annenkov V and Gordon R. A collection of hydrobiont images, http://lin.irk.ru/nanochem/diatom_movies/films.html (2022).

941. Harbich T. Observation of Diatoms: Water surface, https://diatoms.de/en/observation-of-diatoms/movement-on-the-water-surface (2023).

942. Harbich T. Some observations of movements of pennate diatoms in cultures and their possible interpretation. In: Cohn SA, Manoylov KM and Gordon R (eds) *Diatom Gliding Motility [DIGM, Volume in the series: Diatoms: Biology & Applications, series editors: Richard Gordon & Joseph Seckbach]*. Beverly, MA, USA: Wiley-Scrivener, 2021, pp. 1–32.

943. Harbich T. Pattern formation in *Diatoma vulgaris* colonies: Observations and description by a Lindenmayer-system. In: Pappas JL (ed) *The Mathematical Biology of Diatoms [DMTH, Volume in the series: Diatoms: Biology & Applications, series editors: Richard Gordon & Joseph Seckbach]*. Beverly, Massachusetts, USA: Wiley-Scrivener, 2023, pp. 265–290.

944. Harbich T. On the size sequence of diatoms in clonal chains. In: Annenkov V, Seckbach J and Gordon R (eds) *Diatom Morphogenesis [DIMO, Volume in the series: Diatoms: Biology & Applications, series editors: Richard Gordon & Joseph Seckbach]*. Beverly, MA, USA: Wiley-Scrivener, 2021, pp. 69–92.

945. 22nd International Diatom Symposium. Application of new fluorescence-tagged amines and polymers in study of diatom physiology, http://www.lin.irk.ru/nanochem/conf/conf_2_e.html (2012).

946. Ghobara M, Tiffany MA, Gordon R and Reissig L. Diatom pore arrays' periodicities and symmetries in the Euclidean plane: Nature between perfection and imperfection [Chapter 6]. In: Annenkov V, Seckbach J and Gordon R (eds) *Diatom Morphogenesis [DIMO, Volume in the series: Diatoms: Biology & Applications, series editors: Richard Gordon & Joseph Seckbach]*. Beverly, MA, USA: Wiley-Scrivener, 2021, pp. 117–158.

947. Ghobara MM, Vinayak V, Smith DR, Schoefs B, Gebeshuber IC and Gordon R. Diatom frustules as photo-regulators of diatom photobiology In: Yadava RN (ed) *National symposium on Horizons of light in molecules, materials and daily life, December 18–19, 2015, Department of Chemistry (School of Chemical Sciences & Technology), Dr H S Gour Central University, Sagar, MP 470003, India*. Sagar, MP India: Department of Chemistry (School of Chemical Sciences & Technology), Dr. H. S. Gour Central University, 2015.

948. Ghobara M, Gordon R and Reissig L. The mesopores of raphid pennate diatoms: Toward natural controllable mesoporous silica microparticles [Chapter 16]. In: Annenkov V, Seckbach J and Gordon R (eds) *Diatom Morphogenesis [DIMO, Volume in the series: Diatoms: Biology & Applications, series editors: Richard Gordon & Joseph Seckbach]*. Beverly, MA, USA: Wiley-Scrivener, 2020, pp. 383–409.

949. Clark DP and Badea CT. Advances in micro-CT imaging of small animals. *Physica Medica: European Journal of Medical Physics* 2021; 88: 175–192. DOI: 10.1016/j.ejmp.2021.07.005.

950. Beech M, Gordon R and Seckbach J. Terraforming Mars [TMRS, Volume in the series Astrobiology Perspectives on Life of the Universe, Eds. Richard Gordon & Joseph Seckbach]. In: Gordon R and Sechbach J, (eds.). *Astrobiology Perspectives on Life of the Universe*. Beverly, Massachusetts, USA: Wiley-Scrivener, 2021.

951. Rai I, Ahirwar A, Sirotiya V, Gordon R and Vinayak V. Testing diatom growth in Martian environment: A way towards terraforming Mars [Abstract ID Number: 57]. *Wernher von Braun Memorial Symposium,*

Space at the Table: Collaboration, Cooperation, and Inclusion, October 26–28, 2022, The University of Alabama in Huntsville, Huntsville, Alabama. American Astronautical Society, 2022.

952. Rai I, Achankunju J and Gordon R. Terraforming Mars with microalgae, especially diatoms. In: Vinayak V and Gordon R (eds) *Diatom Cultivation for Biofuel, Food and High Value Products [DCUL, Volume in the series: Diatoms: Biology & Applications, series editors: Janice Pappas & Joseph Seckbach, in press]*. Beverly, Massachusetts, USA: Wiley-Scrivener, 2024, pp. in press.

953. Karniadakis G, Beskok A and Aluru N. *Microflows and Nanoflows: Fundamentals and Simulation*. New York: Springer, 2006.

954. Worcester Polytechnic Institute. Ahmet Can Sabuncu, https://www.wpi.edu/people/faculty/acsabuncu (2024).

955. Inhofe M. Ali Beskok, Ph.D., Hunt Institute Senior Fellow, https://blog.smu.edu/huntinstitute/2018/03/27/introducing-ali-beskok-hunt-institute-senior-fellow/ (2018).

956. Wikipedia. Low-density polyethylene, https://en.wikipedia.org/wiki/Low-density_polyethylene (2024).

957. Gordon R and Seckbach J. Preface to *The Science of Algal Fuels: Phycology, Geology, Biophotonics, Genomics, and Nanotechnology*. In: Gordon R and Seckbach J (eds) *The Science of Algal Fuels: Phycology, Geology, Biophotonics, Genomics and Nanotechnology*. Dordrecht: Springer, 2012, pp. xxi-xxii.

958. Encyclopedia Britannica. Green revolution, https://www.britannica.com/event/green-revolution (2024).

959. Wells ML, Potin P, Craigie JS, Raven JA, Merchant SS, Helliwell KE, Smith AG, Camire ME and Brawley SH. Algae as nutritional and functional food sources: Revisiting our understanding. *J Appl Phycol* 2017; 29: 949–982. 20161121. DOI: 10.1007/s10811-016-0974-5.

960. Milko V, Davey E and Fassett C. Indonesia's massive metals build-out is felling the forest for batteries: Indonesia is building out a vast industry for nickel, used for batteries in things like electric cars and cell phones. An analysis of government data reviewed by The AP shows deforestation more than doubles when nickel processing plants come online, https://

apnews.com/article/indonesia-nickel-deforestation-rainforest-mining-tesla-ev-184550cddf1df6aad8e883862ab366df (2024).

961. Sunder M, Acharya N, Nayak S, Gordon R and Mazumder N. Types of x-ray techniques for diatom research [Chapter 8]. In: Mazumder N and Gordon R (eds) *Diatom Microscopy [DIMI, Volume in the series: Diatoms: Biology & Applications, series editors: Richard Gordon & Joseph Seckbach]*. Beverly, MA, USA: Wiley-Scrivener, 2022, pp. 221–236.

962. Alvare G and Gordon R. CT brush and CancerZap!: Two video games for computed tomography dose minimization. *Theor Biol Med Model* 2015; 12: #7. DOI: 10.1186/s12976-015-0003-4.

963. Siebers P and Recker F. Optimierung der medizinischen Ausbildung in Geburtshilfe und Gynäkologie durch Gamification [Optimizing medical education in obstetrics and gynecology through gamification]. *Die Gynäkologie* 2024. DOI: 10.1007/s00129-024-05267-1.

964. Smoukov SK, Seckbach J and Gordon R. Conflicting Models for the Origin of Life [COLF, Volume in the series Astrobiology Perspectives on Life of the Universe, Eds. Richard Gordon & Joseph Seckbach]. Beverly, Massachusetts, USA: Wiley-Scrivener, 2023.

965. Smoukov S, Seckbach J and Gordon R. Preface to Conflicting Models for the Origin of Life. In: Smoukov S, Seckbach J and Gordon R (eds) *Conflicting Models for the Origin of Life [COLF, Volume in the series: Astrobiology Perspectives on Life of the Universe, Series editors: Richard Gordon & Joseph Seckbach, Wiley-Scrivener]*. Beverly, Massachusetts, USA: Wiley-Scrivener, 2023, pp. xix-xxiii.

966. Igamberdiev AU, Beloussov LV and Gordon R. Editorial. Special Issue: Biological Morphogenesis: Theory and Computation [English]. *BioSystems* 2012; 109: 241–242.

967. Portegys T, Pascualy G, Gordon R and Alicea B. Morphozoic: Cellular automata with nested neighborhoods as a novel representation for morphogenesis [Chapter 3]. In: Adamatti DF (ed) *Multi-Agent-Based Simulations Applied to Biological and Environmental Systems*. Hershey, Pennsylvania, USA: IGI Global, 2015, pp. 44–80.

968. Alicea B, Gordon R and Parent J. Embodied cognitive morphogenesis as a route to intelligent systems. *Interface Focus* 2023; 13: #20220067. DOI: 10.1098/rsfs.2022.0067.

969. Sarma G, Lee CW, Portegys T, Ghayoomi V, Alicea B, Cantarelli M, Palyanov A, Khayrulin S, Idili G, Gleeson P, Gordon R, Watts M, Hasani R, Lung D, Currie M, Jacobs T, Gingell S, Gerkin RC and Larson SD. Openworm: Overview and recent advances in integrative biological simulation of *Caenorhabditis elegans*. *Philosophical Transactions of the Royal Society B* 2018; 373B. DOI: 10.1098/rstb.2017.0382.

970. Alicea B, Ahmad B and Richard G. Chapter 8: An ensemble approach to the origin of life. In: Gordon R (ed) *Origin of Life via Archaea: Shaped Droplets to Archaea First, With a Compendium of Archaea Micrographs [OOLA, Volume in the series Astrobiology Perspectives on Life of the Universe, Eds Richard Gordon & Joseph Seckbach, in press]*. Beverly, Massachusetts, USA: Wiley-Scrivener, 2024.

971. Alicea B, Gordon R and Portegys TE. Data-theoretical synthesis of the early developmental process. *Neuroinformatics* 2021; 20: 7–23. DOI: DOI: 10.1007/s12021-020-09508-1.

972. Hartnoll R. In remembrance of Donald Williamson (6 January 1922–29 January 2016), planktologist, carcinologist, and evolutionist: A metamorphosis. *J Crustac Biol* 2016; 36: 408–413. DOI: 10.1163/1937240x-00002430.

973. Wikipedia. Donald I. Williamson, https://en.wikipedia.org/wiki/Donald_I._Williamson (2024).

974. Vickers SE and Williamson DI. Interspecies hybrids. In: Margulis L, Asikainen CA and Krumbein WE (eds) *Chimeras and Consciousness: Evolution of the Sensory Self*. Cambridge: MIT Press, 2011, pp. 183–197.

975. Williamson DI. *Larvae and Evolution: Toward a New Zoology*. New York: Chapman & Hall, 1992.

976. Williamson DI. *The Origins of Larvae*. 2nd ed. Dordrecht, Netherlands: Kluwer Academic Publishers, 2003.

977. Jheeta S, Dominik M, Chatzitheodoridis E, Kotsyurbenko O, Pérez MP, de Farias ST, McGrath K, Rezaei A, Bhatt MC and Gordon R. NoRCEL's Engagement in Africa: The AstroScience Exploration Network (ASEN).

Astrobiology 2023; 8: DOI: 10.1089/ast.2022.0071. DOI: 10.1089/ast.2022.0071.

978. Rai I, Achankunju J, Ghobara M and Vinayak V. Terraforming Mars with diatoms: The perspective of life in a new plane. In: Vinayak V and Gordon R (eds) *Diatom Cultivation for Biofuel, Food and High Value Products [DCUL, Volume in the series: Diatoms: Biology & Applications, series editors: Richard Gordon & Joseph Seckbach]*. Beverly, MA, USA: Wiley-Scrivener, 2023, pp. In press.

979. Sinnott EW. *Plant Morphogenesis*. McGraw-Hill, 1960, p. x+550p. Illus.-x+550p. Illus.

980. Wikipedia. Edmund Ware Sinnott, https://en.wikipedia.org/wiki/Edmund_Ware_Sinnott (2023).

981. Sinnott EW. *The Bridge of Life, From Matter to Spirit*. New York: Simon and Schuster, 1966.

982. Sinnott EW. *Matter, Mind and Man: The Biology of Human Nature*. New York: Atheneum, 1962.

983. Sinnott EW. *Life and Mind*. Antioch Press, 1956.

984. Sinnott EW. *Two Roads to Truth: A Basis for Unity Under the Great Tradition*. Viking Press, 1953.

985. Sinnott EW. *Cell and Psyche; the Biology of Purpose*. New York: Harper, 1950.

986. Johnson BW and Wing BA. Limited Archaean continental emergence reflected in an early Archaean ^{18}O-enriched ocean. *Nat Geosci* 2020; 13: 243–248. DOI: 10.1038/s41561-020-0538-9.

987. Gordon R, Hanczyc MM, Denkov ND, Tiffany MA and Smoukov SK. Emergence of polygonal shapes in oil droplets and living cells: The potential role of tensegrity in the origin of life. In: Gordon R and Sharov AA (eds) *Habitability of the Universe Before Earth [Volume 1 in series: Astrobiology: Exploring Life on Earth and Beyond, eds Pabulo Henrique Rampelotto, Joseph Seckbach & Richard Gordon]*. Amsterdam, Netherlands: Elsevier B.V., 2018, pp. 427–490.

988. Corral P, de la Haba RR, Sanchez-Porro C, Amoozegar MA, Papke RT and Ventosa A. *Halorubrum persicum* sp. nov., an extremely halophilic archaeon isolated from sediment of a hypersaline lake. *Int J Syst Evol Microbiol* 2015; 65: 1770–1778. DOI: 10.1099/ijs.0.000175.

989. Damer B, Newman P, Norkus R, Gordon R and Barbalet T. Cyber-biogenesis and the EvoGrid: a Twenty-First Century grand challenge. In: Seckbach J (ed) *Genesis — In the Beginning: Precursors of Life, Chemical Models and Early Biological Evolution*. Dordrecht: Springer, 2012, pp. 267–288.

990. Wikipedia. David W. Deamer, https://en.wikipedia.org/wiki/David_W._Deamer (2023).

991. Wikipedia. File:Bruce Damer.jpg, https://commons.wikimedia.org/wiki/File:Bruce_Damer.jpg (2023).

992. Barbalet T. Noble ape simulation. *IEEE Comput Graph Appl* 2004; 24: 6–12. DOI: 10.1109/mcg.2004.1274054.

993. Gordon R. When we were all triangles: Shape in the origin of life via abiotic-shaped droplets to living, polygonal Archaea during the Abiocene [Chapter 5]. In: Smoukov S, Seckbach J and Gordon R (eds) *Conflicting Models for the Origin of Life [COLF, Volume in the series: Astrobiology Perspectives on Life of the Universe, Series editors: Richard Gordon & Joseph Seckbach, Wiley-Scrivener]*. Beverly, Massachusetts, USA: Wiley-Scrivener, 2023, pp. 213–262.

994. Gordon R, Raj Vansh Singh S, Katyal K, Gordon NK and Deamer D. The fish ladder toy model for a thermodynamically at equilibrium origin of life in a lipid world in an endoreic lake [Chapter 17]. In: Smoukov SK, Seckbach J and Gordon R (eds) *Conflicting Models for the Origin of Life [COLF, Volume in the series Astrobiology Perspectives on Life of the Universe, Eds Richard Gordon & Joseph Seckbach]*. Beverly, Massachusetts, USA: Wiley-Scrivener, 2023, pp. 425–458.

995. Alicea B, Ahmad B and Gordon R. An ensemble approach to the origin of life [Chapter 8]. In: Gordon R and Seckbach J (eds) *Origin of Life via Archaea: Shaped Droplets to Archaea First, With a Compendium of Archaea Micrographs [OOLA, Volume in the series Astrobiology Perspectives on Life of the Universe]*. Beverly, Massachusetts, USA: Wiley-Scrivener, 2024, pp. 285–340.

996. Gordon R. The chromolinker hypothesis: Are eukaryotic genomes also circular? *BioSystems* 2024; 244: #105280. DOI: https://doi.org/10.1016/j.biosystems.2024.105280.

997. Glanz J. Force-carrying web pervades living cell. *Science* 1997; 276: 678–679.

998. Ball P. *How Life Works: A User's Guide to the New Biology*. University of Chicago Press, 2023.

999. Moon RJ. Amplifying and intensifying the fluoroscopic image by means of a scanning x-ray tube. *Science* 1950; 112: 389–395.

1000. Annenkov VV, Zelinskiy SN, Pal'shin VA, Larina LI and Danilovtseva EN. Coumarin based fluorescent dye for monitoring of siliceous structures in living organisms. *Dyes Pigment* 2019; 160: 336–343. DOI: 10.1016/j.dyepig.2018.08.020.

1001. Pickett-Heaps JD, Spurck T, Wetherbee R, Cohn S and Pickett-Heaps J. *Diatoms: Life in Glass Houses [DVD]*. Cytographics, 2003.

1002. Gordon R. Response to: Woloshin, S. & L.M. Schwartz (2012). How a charity oversells mammography. BMJ 345(1), #e5132. *BMJ* 2012; 345: http://www.bmj.com/content/345/bmj.e5132?tab=responses. DOI: 10.1136/bmj.e5132.

1003. Gordon R and Carron I. Steered microbeams and other active optics as compressive sensing, http://nuit-blanche.blogspot.com/2010/02/cs-blind-compressed-sensing-some-more.html (2010).

1004. Verschraegen C, Vinh-Hung V, Cserni G, Gordon R, Royce ME, Vlastos G, Tai P and Storme G. Modeling the effect of tumor size in early breast cancer. *Ann Surg* 2005; 241: 309–318.

1005. Chapman J-A, Gordon R, Link MA and Fish EB. Infiltrating breast carcinoma smaller than 0.5 centimeters. Is lymph node dissection necessary? *Cancer* 1999; 86: 2186–2188.

1006. Sun J, Chapman JA, Gordon R, Sivaramakrishna R, Link MA and Fish EB. Survival from primary breast cancer after routine clinical use of mammography. *The Breast Journal* 2002; 8: 199–208.

1007. Sun J, Chapman JA, Gordon R, Sivaramakrishna R, Link MA and Fish EB. Survival from primary breast cancer by tumour size for the age groups with different screening guidelines. *Breast Cancer Res Treatment* 1998; 50: 281.

1008. Vinh-Hung V and Gordon R. Quantitative target sizes for breast tumor detection prior to metastasis: A prerequisite to rational design of 4D scanners for breast screening. *Technol Cancer Res Treat* 2005; 4: 11–21. 2005/01/15. DOI: d=3022&c=4170&p=12635&do=detail [pii].

1009. Sivaramakrishna R and Gordon R. Detection of breast cancer at a smaller size can reduce the likelihood of metastatic spread: A quantitative analysis. *Acad Radiol* 1997; 4: 8–12.

1010. Carron I and Gordon R. CT Brush and CancerZap!: Two video games for computed tomography dose minimization, http://nuit-blanche. blogspot.ca/2015/06/ct-brush-and-cancerzap-two-video-games.html?utm_source=feedburner&utm_medium=email&utm_campaign=-Feed:+blogspot/vhVI+(Nuit+Blanche) (2015).

1011. Merz CR, Liu Y and Weisberg RH. Sea Surface Current Mapping with HF Radar — a Primer. *University of South Florida, Marine Science Faculty Publications* 2021: #2538.

1012. Gordon R and Brodland GW. The formation of the brain and spina bifida. *SB Podium, Newsletter of the Spina Bifida Association of Canada* 1987; 2: 6–8.

1013. Haeckel EHPA. *Kunstformen der Natur.* Leipzig, Germany: Bibliographisches Institut, 1904.

1014. Haeckel E. *The Love Letters of Ernst Haeckel.* London, UK: Methuen & Co Ltd., 1930.

1015. Gordon R and Seckbach J. The Science of Algal Fuels: Phycology, Geology, Biophotonics, Genomics and Nanotechnology [ALFU]. Dordrecht: Springer, 2012.

1016. Goessling JW, Serôdio J and Lavaud J. Diatom Photosynthesis: From Primary Production to High-Value Molecules [DPHO, Volume in the series: Diatoms: Biology & Applications, series editors: Richard Gordon & Joseph Seckbach]. In: Gordon R and Seckbach J, (eds.). *Diatoms: Biology & Applications.* Beverly, MA, USA: Wiley-Scrivener, 2024.

1017. Beech M. Geoengineering and Climate Change: Methods, Risks, and Governance [GENG, Volume in the series Astrobiology Perspectives on Life of the Universe, in press Series Editors: Richard Gordon & Joseph Seckbach]. In: Gordon R and Seckbach J, (eds.). Beverly, Massachusetts, USA: Wiley-Scrivener, 2024.

1018. Seckbach J and Stan-Lotter H. Extremophiles as Astrobiological Models [EXAM, Volume in the series Astrobiology Perspectives on Life of the

Universe, Eds. Richard Gordon & Joseph Seckbach]. In: Gordon R and Sechbach J, (eds.). *Astrobiology Perspectives on Life of the Universe*. Beverly, Massachusetts, USA: Wiley-Scrivener, 2021.

1019. Chon Torres OA, Peters T, Seckbach J and Gordon R. Astrobiology: Science, Ethics and Public Policy[ABET, volume in the series: Astrobiology Perspectives on Life of the Universe (Series editors: Richard Gordon & Joseph Seckbach)]. Beverly, Massachusetts, USA: Wiley-Scrivener, 2021.

1020. Berea A. Technosignatures: Searching for and Communicating with Intelligence in Our Universe [TECH, Volume in the series Astrobiology Perspectives on Life of the Universe, Eds. Richard Gordon & Joseph Seckbach]. Beverly, Massachusetts, USA: Wiley-Scrivener, 2022.

1021. Tinnenbroek A. Frank Sinatra: I Did It My Way, https://www.youtube.com/watch?v=9SXWX6qg0y4 (2013).

1022. School of the Art Institute of Chicago. Ox-Bow School of Art, https://www.saic.edu/academics/campus-study/ox-bow-school-art (2024).

1023. Seckbach R. File:Joseph Seckbach.jpg, https://commons.wikimedia.org/wiki/File:Joseph_Seckbach.jpg (2012).